Novel Magnetic Properties in Curved Geometries

Novel Magnetic Properties in Curved Geometries

Editor

Cristina Bran

MDPI • Basel • Beijing • Wuhan • Barcelona • Belgrade • Manchester • Tokyo • Cluj • Tianjin

Editor
Cristina Bran
Materials for Information
Technologies/Group of
Nanomagnetism and
Magnetization Processes
Instituto de Ciencia de
Materiales de Madrid
(ICMM-CSIC)
Madrid
Spain

Editorial Office
MDPI
St. Alban-Anlage 66
4052 Basel, Switzerland

This is a reprint of articles from the Special Issue published online in the open access journal *Nanomaterials* (ISSN 2079-4991) (available at: www.mdpi.com/journal/nanomaterials/special_issues/Magnetic_Geometries).

For citation purposes, cite each article independently as indicated on the article page online and as indicated below:

LastName, A.A.; LastName, B.B.; LastName, C.C. Article Title. *Journal Name* **Year**, *Volume Number*, Page Range.

ISBN 978-3-0365-3910-2 (Hbk)
ISBN 978-3-0365-3909-6 (PDF)

Contents

About the Editor

Cristina Bran

Cristina Bran completed her Ph.D. in physics in 2010 at IFW Dresden for study on magnetization processes of antiferromagnetic coupled multilayers. She then joined the Division of Material Physics at Uppsala University as a postdoc working on synthesis and magnetic characterization of tetragonally strained superlattices. Currently, she is in charge of Electrochemistry/ Nanofabrication Laboratory at Institute of Materials Science of Madrid dedicated to synthesis and fabrication of multifunctional magnetic nanostructures. Her current research interests focus on the nanomagnetism of 3D cylindrical structures, including both materials preparation and nanoscale characterization, for potential applications in spintronic, biomedicine and energy transfer applications.

nanomaterials

Editorial

Novel Magnetic Properties in Curved Geometries

Cristina Bran 

Instituto de Ciencia de Materiales de Madrid (ICMM-CSIC), 28049 Madrid, Spain; cristina.bran@icmm.csic.es

Citation: Bran, C. Novel Magnetic Properties in Curved Geometries. *Nanomaterials* **2022**, *12*, 1175. https://doi.org/10.3390/nano12071175

Received: 14 March 2022
Accepted: 25 March 2022
Published: 1 April 2022

Publisher's Note: MDPI stays neutral with regard to jurisdictional claims in published maps and institutional affiliations.

The expanding of planar magnetic structures into three dimensions (3D) creates the possibility of tuning the conventional magnetic textures or producing novel effects and functionalities by tailoring their curvature [1]. The effect of curvature in the magnetic systems is an emerging topic of research because of the novel magnetization textures and dynamics related to topology as compared to planar geometries. The interesting capabilities of 3D magnetic structures with curved geometry come in the first place from the enormous increase in the active surface for storage, as well as from their versatile geometrical forms, resulting in complex 3D magnetic configurations with great potential for manipulation [2].

One of the most promising magnetic structures with curved geometry are cylindrical magnetic wires, as they offer multifunctional responses to electric/ magnetic fields, electric current, mechanical stress, or thermal gradients and thus can be used for the interconversion between different functionalities.

Their circular symmetry determines multiple topologically non-trivial magnetization structures, such as Bloch-point domain walls, which are faster and more stable than domain walls in planar structures, or helical magnetic configurations, vortices, and skyrmion tubes [3,4].

Exploiting these unusual three-dimensional magnetic configurations and their dynamics can lead to advanced applications such as new-generation spintronic-based magnetic recording, bio-magnetics, robotics, sensors, and actuators devices [5].

This Special Issue of *Nanomaterials* covers the recent advancements in the fabrication, characterization, and potential technological applications of magnetic wires, as single magnetic structures or as part of 3D ordered architectures, in magnetic sensors and data storage, microwave devices, or thermomagneto-electric devices.

The first review [6] of the Special Issue presents many examples of how the magnetic structure can be tailored and controlled in cylindrical nanowires by geometry and magnetocrystalline anisotropy or manipulated by magnetic fields and spin-polarized currents. The nanowires are prepared by electrochemical methods into the porous pores of alumina templates, allowing the fabrication of magnetic nanowires with precise control over geometry, morphology, and composition. The diameter modulations change the typical single domain state present in crystallographic cubic nanowires, providing the possibility to confine or pin circular domains or domain walls in each segment. The control and stabilization of domains and domain walls in cylindrical wires have been achieved in multi-segmented structures by alternating magnetic segments of different magnetic properties (producing alternative anisotropy) or with non-magnetic layers. The results point out the relevance of the geometry and magnetocrystalline anisotropy to promote the occurrence of stable magneto-chiral structures and provide further information for the design of cylindrical nanowires for multiple applications [6].

Similarly, by playing with the architecture of cylindrical magnetic nanostructures, J. Garcia el al. [7] presents a magnetic study on core/shell nanostructures formed by a nanowire nucleus ($Fe_{56}Co_{44}$), grown by electrodeposition and coated by a non-magnetic SiO_2 layer coaxially surrounded by a magnetic Fe_3O_4 nanotubular coating both fabricated by means of the Atomic Layer Deposition technique. The magnetic reversal studied by First Order Reversal Curve methodology reveals a two-step magnetization reversal of the core/shell bi-magnetic nanostructure. These results are also confirmed by the hysteresi loops of individual core/shell nanostructures measured by Kerr effect-based magnetomete

Gomes et al. [8], present an interconnected 3D nanowire system fabricated by direc electrodeposition in track-etched polymer templates with crossed nano-channels. Thi method allows the fabrication of crossed single element and multilayered nanowires wit controlled morphology and material composition. The interconnected nanowire network exhibit high room-temperature Seebeck coefficient and power factors, making them promi ing candidates for the next generation of flexible and lightweight thermoelectric devices [8

An alternative method to produce curved 3D nanostructures with defined and con plex geometries is presented in the review paper by C. Magen et al. [9]. Focused electro beam induced deposition (FEBID) is a single-step additive nanolithography techniqu based on the local decomposition of the molecules of an organometallic precursor ga adsorbed on the surface of a substrate, thus producing a solid deposit. The review present different possibilities for engineering the geometrical, compositional, and magnetic prope ties of 3D ferromagnetic nanowires grown by FEBID, focusing on the fine-tuning of FEBII growth parameters to modify the composition and dimensions, the growth of core-shel heterostructures, and the purification by thermal annealing.

An important part of this Special Issue deals with the magnetic microwires. Th peculiarities of microwires circular symmetry, outstanding magnetic properties, and th cheap and accessible fabrication method makes them attractive from sensing application

The review article by Chizhik et al. [10] summarizes the results of the magneti magneto-electric, and magneto-optical studies of a series of glass-covered microwires. Th experimental data are complemented by micromagnetic simulations. Different tools fo controlling the reversible and irreversible transformations of the magnetic structure o glass-covered microwires are evidenced. The irreversible tools are the selection of th chemical composition, geometric ratio, and the stress-annealing, while a combination o magnetic fields and mechanical stresses is needed for the reversible tuning. The study i also focused on the giant magnetoimpedance effect and the velocity of the domain wall propagation important for the technological applications.

Alekhina et al. [11] propose an indirect method for the estimation of micromagneti structure in glass-coated microwires. The cross-sectional permeability distribution in th microwires was obtained from impedance measurements at different frequencies. Thi distribution enables estimation of the prevailing anisotropy in the local region of the wir cross-section. The results obtained were compared with the findings of magnetostati measurements and remanent state analysis. The advantages and limitations of the method are also discussed.

The Magnetoimpedance (MI) of Co-based microwires with an amorphous and pa tially crystalline state are investigated at elevated frequencies by Alam et al. [12]. Tw mechanisms of MI sensitivity related to the DC magnetization re-orientation and AC pe meability dispersion are discussed. Remarkable sensitivity of impedance changes wit respect to applied tensile stress at GHz frequencies are obtained in partially crystallin wires subjected to current annealing. Increasing the annealing current enhanced the axi easy anisotropy of a magnetoelastic origin, which made it possible to increase the frequenc of large-stress MI.

This collection of interesting research papers shows some of the capabilities an challenges in the preparation, characterization, and possible applications of magnetic wire with cylindrical geometry.

Both from an experimental and micromagnetic modelling point of view, this Speci Issue present interesting ideas, showing the prospects and exciting developments in th field of cylindrical magnetic wires.

Finally, I would like to express my gratitude to all the authors for their valuab contribution to this Special Issue.

Funding: This research received no external funding.

Institutional Review Board Statement: Not applicable.

Informed Consent Statement: Not applicable.

Data Availability Statement: Not applicable.

Acknowledgments: I acknowledge the reviewers for enhancing the quality and impact of all submitted papers.

Conflicts of Interest: The author declares no conflict of interest.

References

Streubel, R.; Fischer, P.; Kronast, F.; Kravchuk, V.P.; Sheka, D.D.; Gaididei, Y.; Schmidt, O.G.; Makarov, D. Magnetism in curved geometries. *J. Phys. D Appl. Phys.* **2012**, *49*, 363001. [CrossRef]

Fernández-Pacheco, A.; Streubel, R.; Fruchart, O.; Hertel, R.; Fischer, P.; Cowburn, R.P. Three-dimensional nanomagnetism. *Nat. Commun.* **2017**, *8*, 15756. [CrossRef] [PubMed]

Sanz-Hernández, D.; Hierro-Rodriguez, A.; Donnelly, C.; Pablo-Navarro, J.; Sorrentino, A.; Pereiro, E.; Magén, C.; McVitie, S.; de Teresa, J.M.; Ferrer, S.; et al. Artificial Double-Helix for Geometrical Control of Magnetic Chirality. *ACS Nano* **2020**, *14*, 8084. [CrossRef] [PubMed]

Charilaou, M.; Braun, H.B.; Löffler, J.F. Monopole-Induced Emergent Electric Fields in Ferromagnetic Nanowires. *Phys. Rev. Lett.* **2018**, *121*, 97202. [CrossRef] [PubMed]

Stano, M.; Fruchart, O. Chapter 3—Magnetic Nanowires and Nanotubes. In *Handbook of Magnetic Materials*; Elsevier: Amsterdam, The Netherlands, 2018; Volume 27, pp. 155–267. [CrossRef]

Bran, C.; Fernandez-Roldan, J.A.; del Real, R.P.; Asenjo, A.; Chubykalo-Fesenko, O.; Vazquez, M. Magnetic Configurations in Modulated Cylindrical Nanowires. *Nanomaterials* **2021**, *11*, 600. [CrossRef] [PubMed]

García, J.; Manterola, A.M.; Méndez, M.; Fernández-Roldán, J.A.; Vega, V.; González, S.; Prida, V.M. Magnetization Reversal Process and Magnetostatic Interactions in $Fe_{56}Co_{44}/SiO_2/Fe_3O_4$ Core/Shell Ferromagnetic Nanowires with Non-Magnetic Interlayer. *Nanomaterials* **2021**, *11*, 2282. [CrossRef] [PubMed]

Da Câmara Santa Clara Gomes, T.; Marchal, N.; Abreu Araujo, F.; Piraux, L. Spin Caloritronics in 3D Interconnected Nanowire Networks. *Nanomaterials* **2020**, *10*, 2092. [CrossRef] [PubMed]

Magén, C.; Pablo-Navarro, J.; De Teresa, J.M. Focused-Electron-Beam Engineering of 3D Magnetic Nanowires. *Nanomaterials* **2021**, *11*, 402. [CrossRef] [PubMed]

Chizhik, A.; Gonzalez, J.; Zhukov, A.; Gawronski, P.; Ipatov, M.; Corte-León, P.; Blanco, J.M.; Zhukova, V. Reversible and Non-Reversible Transformation of Magnetic Structure in Amorphous Microwires. *Nanomaterials* **2020**, *10*, 1450. [CrossRef] [PubMed]

Alekhina, I.; Kolesnikova, V.; Rodionov, V.; Andreev, N.; Panina, L.; Rodionova, V.; Perov, N. An Indirect Method of Micromagnetic Structure Estimation in Microwires. *Nanomaterials* **2021**, *11*, 274. [CrossRef] [PubMed]

Alam, J.; Nematov, M.; Yudanov, N.; Podgornaya, S.; Panina, L. High-Frequency Magnetoimpedance (MI) and Stress-MI in Amorphous Microwires with Different Anisotropies. *Nanomaterials* **2021**, *11*, 1208. [CrossRef] [PubMed]

 *nanomaterials*

Review

Magnetic Configurations in Modulated Cylindrical Nanowires

Cristina Bran [1,*], Jose Angel Fernandez-Roldan [1,2], Rafael P. del Real [1], Agustina Asenjo [1], Oksana Chubykalo-Fesenko [1] and Manuel Vazquez [1]

1   Instituto de Ciencia de Materiales de Madrid, CSIC, 28049 Madrid, Spain; fernandezroljose@uniovi.es (J.A.F.-R.); rafael.perez@icmm.csic.es (R.P.d.R.); aasenjo@icmm.csic.es (A.A.); oksana@icmm.csic.es (O.C.-F.); mvazquez@icmm.csic.es (M.V.)
2   Department of Physics, University of Oviedo, 33007 Oviedo, Spain
*   Correspondence: cristina.bran@icmm.csic.es

Abstract: Cylindrical magnetic nanowires show great potential for 3D applications such as magnetic recording, shift registers, and logic gates, as well as in sensing architectures or biomedicine. Their cylindrical geometry leads to interesting properties of the local domain structure, leading to multifunctional responses to magnetic fields and electric currents, mechanical stresses, or thermal gradients. This review article is summarizing the work carried out in our group on the fabrication and magnetic characterization of cylindrical magnetic nanowires with modulated geometry and anisotropy. The nanowires are prepared by electrochemical methods allowing the fabrication of magnetic nanowires with precise control over geometry, morphology, and composition. Different routes to control the magnetization configuration and its dynamics through the geometry and magnetocrystalline anisotropy are presented. The diameter modulations change the typical single domain state present in cubic nanowires, providing the possibility to confine or pin circular domains or domain walls in each segment. The control and stabilization of domains and domain walls in cylindrical wires have been achieved in multisegmented structures by alternating magnetic segments of different magnetic properties (producing alternative anisotropy) or with non-magnetic layers. The results point out the relevance of the geometry and magnetocrystalline anisotropy to promote the occurrence of stable magnetochiral structures and provide further information for the design of cylindrical nanowires for multiple applications.

Keywords: cylindrical magnetic nanowires; magnetocrystalline anisotropy; magnetochiral configurations; micromagnetic modeling

Citation: Bran, C.; Fernandez-Roldan, A.; del Real, R.P.; Asenjo, A.; Chubykalo-Fesenko, O.; Vazquez, M. Magnetic Configurations in Modulated Cylindrical Nanowires. *Nanomaterials* **2021**, *11*, 600. https://doi.org/10.3390/nano11030600

Academic Editor: Julian Maria Gonzalez Estevez

Received: 28 January 2021
Accepted: 21 February 2021
Published: 28 February 2021

Publisher's Note: MDPI stays neutral with regard to jurisdictional claims in published maps and institutional affiliations.

1. Introduction

The increasing interest in nanomaterials with curved geometry lies in the novel magnetic phenomena observed in those magnetic systems [1]. This can lead to multiple applications which are already being developed currently or are quite promising in the near future. They include magnetochiral phenomena (a consequence of curvature) and other novel effects that open new perspectives not only from fundamental aspects but also in advanced technologies [2,3]. The scientific and technological exploration of three-dimensional magnetic nanostructures is an emerging research field that opens the path to exciting novel physical phenomena, originating from the increased complexity in spin textures, topology, and frustration in three dimensions [4].

A particular case is that of nanowires with a circular cross-section. Among other possibilities, the electrochemical route to fabricate cylindrical nanowires inside ordered porous templates is a less-expensive method that offers wide versatility. Ordered arrays of cylindrical nanowires have attracted much interest due to their broad range of applications that go from 3D magnetic information and logic devices, to advanced sensors based on magnetotransport and magnetomechanical responses, spin-caloritronics, microwave, and magnonics, or as novel permanent magnets [5,6]. Advances in smart electronics, robotics, and virtual reality demand electronic skins with both tactile and touchless perceptions for the manipulation of real and virtual objects; here magnetic microelectromechanical systems based on cylindrical nanowires can be used to transduce both tactile and touchless sensing via magnetic fields [7,8]. More recently individual nanowires have been proposed, after being properly functionalized, for biomedical applications in drug delivery and oncological applications, or for nanorobots and swimming nano & microdevices conducted by applied magnetic fields [9,10].

However, most applications are based on the magnetic behavior of individual cylindrical nanowires. The magnetic response of these nanowires can be firstly tailored through their composition (e.g., based on Fe, Co, and Ni as single magnetic elements and their alloys) that determines the cubic or hexagonal crystal symmetry and consequently the magnetocrystalline anisotropy. That anisotropy together with the shape anisotropy finally determines the stable magnetic configurations. In fact, magnetic properties, as domain structure and specific remagnetization processes, are in the origin of most relevant techno logical applications.

Complex magnetic configurations are either observed experimentally or/and predicted by micromagnetic modeling. Configurations with magnetochiral components are promoted by cylindrical shapes and include vortex domains with complex 3D transitions between them, helical structures, or skyrmion tubes. The remanent stable magnetic domains are determined by the balance between magnetocrystalline and shape anisotropies. Thus, geometry in the relationship between diameter and length in multi segmented nanowires plays an essential role. Similarly, modulations in composition (e.g either ferromagnetic/ferromagnetic segments with differential magnetic characteristic or ferromagnetic/non-ferromagnetic metal segments, imposing non-magnetic barriers significantly influence the domain structure.

Several advanced experimental techniques have been successfully employed to characterize the magnetic state of individual nanowires. That includes magnetic measurement by Magnetic Force Microscopy (MFM) and Magneto-Optical Kerr Effect (MOKE) sensitive to surface magnetism. Other techniques such as X-ray Magnetic Circular Dichroism combined with Photoemission Electron Microscopy (XMCD-PEEM) allow including information of internal configuration. The distribution of magnetic fields has been successfully achieved by advanced electron holography techniques while tomography methods are emerging to visualize their 3D structure [11]. On the other hand, micromagnetic simulations are required to complement and fully understand the complex magnetic structures in cylindrical nanowires.

As for the reversal process, owing to the large length to the diameter aspect ratio the magnetization reversal proceeds directionally along the nanowire when the external magnetic field is applied parallel to its axis. Thus, the reversal proceeds overall from one end to the other via the propagation of domain walls. However, most interestingly point-like singularities commonly labeled as Bloch points, are typically in the middle of domain walls separating axially magnetized domains [12].

The aim of this work is to provide a review of the most relevant results obtained by our group on the fabrication, characterization, and properties of ferromagnetic modulated nanowires with tailored geometry and anisotropy. The review is organized as follows: first, the synthesis and fabrication of high-quality nanowires with tailored geometry and magnetic anisotropy are introduced in Section 2. The experimental and modeling results of modulated nanowires with low and high magnetocrystalline anisotropy are presented in Sections 3 and 4, respectively. Finally, Section 5 presents the discussion and conclusions.

2. Materials and Methods

2.1. Alumina Substrates

Anodic aluminum oxide (AAO) templates with pores with diameters between 100–200 nm are prepared by anodization processes from high purity aluminum disks foils [13–18]. The disks are degreased by sonication in acetone (for 10 min) and ethanol baths (for 15 min). Before anodization, the Al discs are electropolished in a mixed solution of $HClO_4$:C_2H_5OH = 1:3 (*v*/*v*) under a potential of 20 V for 3–5 min. In order to fabricate the templates with pores of uniform diameter over 100 nm, the Al disks are anodized by hard anodization process in oxalic acid 0.3 M solution with 5% of ethanol in volume, at 0–1 °C, following the steps illustrated in Figure 1a. First, a constant voltage of 80 V is applied for 400–600 s (step 1) to produce a protective aluminum oxide layer at the surface of the disc which avoids breaking or burning effects during the second and third steps. In the second step, the voltage is steadily increased (0.07–0.08 V/s) until the highest applied voltage of 120–140 V is reached and kept constant for 3600 s (step 3). The resulting pores with diameters of about 120–130 nm and lengths of about 60 µm (Figure 1c) are distributed in a hexagonal order (Figure 1d) [19,20].

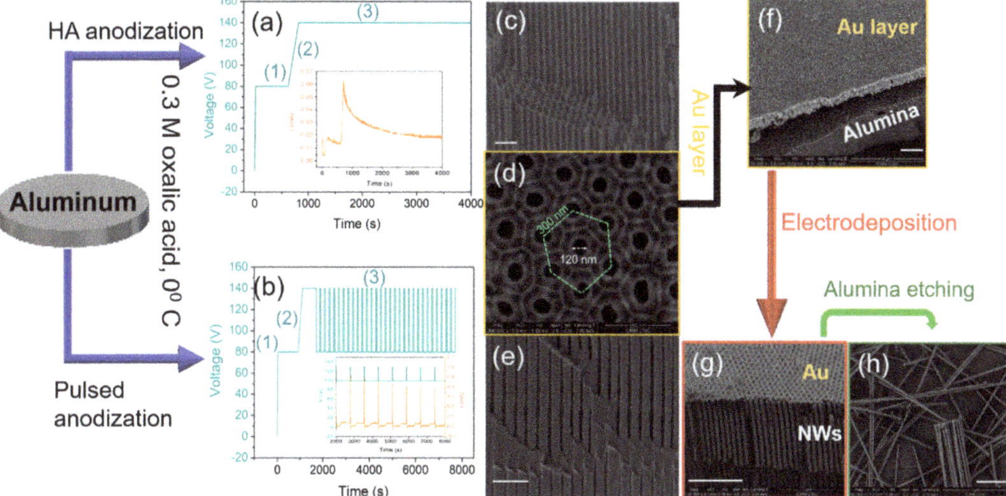

Figure 1. Illustration of the fabrication process. (**a**) Voltage–time transients of hard anodization. The inset shows the current during the anodization, (**b**) Voltage–time transients of pulsed hard anodization, including the first anodization step (1), ramping of voltage (2), and the applied pulses (3). The inset shows a close look of both applied voltage and current transients, during the pulsed anodization, (**c**) Cross-section Scanning Electron Microscopy (SEM) image of alumina templates prepared by hard anodization with pores of uniform diameter of 120 nm, (**d**) Top view SEM image of the bottom side of alumina template presented in (**c**,**e**) Cross-section SEM image of alumina template prepared by pulsed anodization with modulated pores, (**f**) SEM image of an Au layer deposited on the bottom of alumina template, (**g**) cylindrical nanowires deposited by electrodeposition inside the alumina pores, (**h**) individual nanowires released from alumina templates by chemical etching. The scale bar in (**c**,**e**–**h**) is 1 µm.

The modulated pores are fabricated by pulsed anodization [21,22] using the same acidic electrolyte as for the pores with uniform diameter, where step (3) in Figure 1a is replaced by voltage pulses (Figure 1b). Depending on the anodization parameters (voltage pulses, time) different types of modulations can be obtained. The modulated AAO pores presented in Figure 1e are obtained by applying hard anodization pulses of 100 V and 130 V for 100 s and 5 s, respectively.

An alternative method used in our laboratories for obtaining AAO pores (straight and modulated) with smaller diameters involves the use of a sulfuric acid solution.

The pulse anodization of aluminum can also be done by replacing the oxide electrolyte with H_2SO_4 electrolyte. Periodic pulses, alternating a low and high potential pulse are applied, where the duration of each pulse determines the length of anodized segments at the given applied potential. The modulation of nanopores has been achieved by two-step anodization of aluminum discs in a 0.3 M H_2SO_4 at 0 °C temperature. The first anodizing step was performed at a constant potential of 25 V for 16 h [23,24].

After chemical etching of the anodized section, the modulated nanopores are synthesized in the 2nd anodization step by periodically applying pulses of 25 V—mild anodization (MA) pulses, and 35 V—hard anodization pulses (HA) (Figure 2a). The geometry of the nanochannels is controlled not only by the anodization time, voltage, or bath temperature but also by the shape of the HA pulses. The applied voltage pulses with exponential (Figure 2a$_1$) and square (Figure 2a$_2$) shapes produce slightly different modulation of the nanochannels (Figure 2a,b). The resulting cylindrical modulated pores are formed by segments with diameters of around 22 and 35 nm, while the center-to-center inter-wire distance is kept constant at 65 nm.

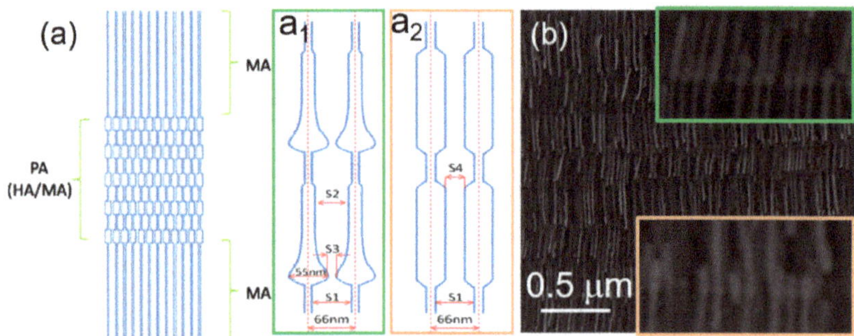

Figure 2. (**a**) Schematic illustration of nanowire geometrical features using modulated diameter within exponential (**a₁**) and squared (**a₂**) pulses template (Adapted with permission from ref. [2]. Copyright 2014 IOP Publishing Ltd.), (**b**) cross-section SEM image of FeCo nanowires with modulated pore diameter formed by pulse anodization in H_2SO_4 electrolyte. The insets show the slight change in the geometry determined by the shape of the voltage pulse.

2.2. Samples Deposition

After the anodization process, the remaining aluminum substrate is chemically etched by a mixed solution of $CuCl_2 \cdot 2H_2O$ and HCl. The alumina barrier layer is removed and the pores are enlarged using a H_3PO_4 solution (5 wt. %). Before depositing the nanowire inside the pores, an Au layer is sputtered on the backside of the AAO template (Figure 1) to serve as a working electrode for electrodeposition. The magnetic nanowires (Figure 1) can be released from the AAO template by using a mixed solution of CrO_3 and H_3PO_4 (Figure 1h).

The magnetic nanowires are grown mainly by potentiostatic electrodeposition into the pores of AAO templates. The deposition is done at suitable potentials in a three-electrode electrochemical cell equipped with an Ag/AgCl reference electrode, a Pt mesh counter and an Au layer sputtered on the backside of the AAO template, acting as a working electrode. The technique allows controlling the composition of the deposited material, from a single element to alloys or multi-segmented nanowires [25–29]. The electrolytes and parameters used in electrodeposition are presented in Table 1.

Following the procedure presented in Section 2.1, four types of nanopores were obtained. By filling them by electrodeposition, magnetic nanowires mirroring the shape of the pores were produced.

Figure 3 presents SEM cross-section images of (a) FeCo nanowires with a uniform diameter of 120 nm (b) bamboo-type FeCoCu nanowires (c) FeCoCu modulated nanowire with alternating segments of 110 and 130 nm in diameter, and (d) Ni nanowires with notches along their length.

Table 1. Materials and parameters used in electrodeposition.

Material	Electrolyte	Voltage
(1) $Fe_{30}Co_{65}Cu_5$ [24]	0.05 M $FeSO_4 \cdot 7H_2O$, 0.12 M $CoSO_4 \cdot 7H_2O$, 0.16 M H_3BO_3, 0.01 M $CuSO_4 \cdot 5H_2O$, 0.06 M $C_6H_8O_6$	−1.8
(2) $Fe_{50}Co_{50}$ [24]	0.08 M $CoSO_4 \cdot 7H_2O$, 0.08 M $FeSO_4 \cdot 7H_2O$, 0.16 M H_3BO_3, 0.06 M $C_6H_8O_6$	−1.8
(3) Ni	0.76 M $NiSO_4 \cdot 6H_2O$, 0.17 M $NiCl_2 \cdot 6H_2O$, 0.65 M H_3BO_3	−1.0
(4) $Co_{65}Ni_{35}$ [20]	0.09 M $CoSO_4 \cdot 7H_2O$ + 0.063 M $CoCl_2 \cdot 6H_2O$ +0.095 M $NiSO_4 \cdot 7H_2O$ + 0.084 M $NiCl_2 \cdot 6H_2O$ + 0.32 M H_3BO_3	−1.1
(5) $Co_{85}Ni_{15}$ [20]	0.12 M $CoSO_4 \cdot 7H_2O$ + 0.084 M $CoCl_2 \cdot 6H_2O$ + 0.064 M $NiSO_4 \cdot 7H_2O$ + 0.063 M $NiCl_2 \cdot 6H_2O$ + 0.32 M H_3BO_3	−1.2

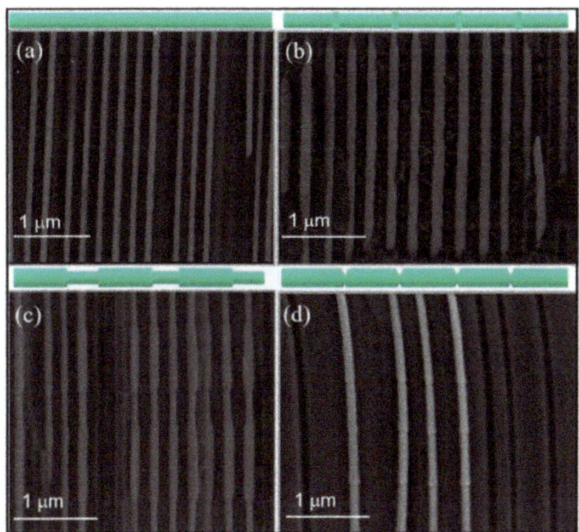

Figure 3. SEM lateral view images of (**a**) uniform (**b**) bamboo-type (**c**) modulated in diameter and (**d**) notched nanowires.

2.3. Characterization Methods

Magneto-Optical Kerr Effect (MOKE). The single nanowires have been measured by a Kerr effect magnetometer NanoMOKE TM 2 from Durham Magneto Optics Ltd. (Durham, UK) under a maximum applied field of ±500 Oe. Each hysteresis loop measured by MOKE is the result of 1000 averaged loops. Once the nanowires are dispersed on a silicon substrate, a scan of the surface was carried out by SEM in order to ensure that the MOKE measurements are focused on individual wires and not on several at once.

Magnetic Force Microscopy (MFM). MFM measurements were performed in a Cervantes system from Nanotec Electrónica (Madrid, Spain). The use of amplitude modulation (AM) and the two-pass modes with a phase-locked loop (PLL) enabled tracking the resonance frequency of the oscillating cantilevers. BudgetSensors Multi75M (BudgetSensors, Sofia, Bulgaria) and Nanosensors PPP-MFMR (NANOSENSORS, Neuchatel, Switzerland) probes were used in these experiments. Parallel topographic images were taken to check the diameter modulation periodicity.

X-ray Magnetic Circular Dichroism combined with Photoemission Electron Microscopy (XMCD-PEEM). XMCD-PEEM measurements were performed at the CIRCE beamline of the ALBA Synchrotron Facility (Barcelona, Spain) using an ELMITEC LEEM III (Clausthal-Zellerfeld, Germany) instrument with an energy analyzer [30]. The samples were illuminated with circularly polarized X-rays at a grazing angle of 16° with respect to the surface, at the resonant L3 absorption edges of Fe (708 eV), Co (778 eV), and Ni (851 eV) for the Fe, CoCu, CoNi, and Ni wires, respectively. The emitted photoelectrons (low energy secondary electron with ca. 1eV kinetic energy) used to form the surface image are proportional to the X-ray absorption coefficient and thus the element-specific magnetic domain configuration is given by the pixel-wise asymmetry of two PEEM images sequentially recorded with left and right-handed circular polarization [31].

In this arrangement of the setup (Figure 4a), an amount of X-ray photons is transmitted through the nanowire, generating photoemission from the Si substrate. Since the transmitted intensity depends on the relative alignment of the nanowire magnetization and the X-ray helicity, the photoemission from the substrate in the area shadowed by the nanowire does as well. Therefore, by analyzing the circular dichroic or pseudo-magnetic contrast formed in transmission in the shadow area, information about the magnetization configuration in the bulk of the wire can be obtained [21,32]. Notice that dark contrast in the transmission is equivalent to bright indirect photoemission since the absorbed and transmitted X-rays are complementary. XMCD-PEEM thus offers the possibility to obtain both the magnetic structure of the surface and the core of the cylindrical structure (Figure 4b). The projection of the local magnetization on the photon propagation vector is determined, the domains with magnetic moments parallel or antiparallel to the X-ray polarization vector appear bright or dark in the XMCD image while domains with magnetic moments at a different angle have an intermediate grey contrast [20,21,29].

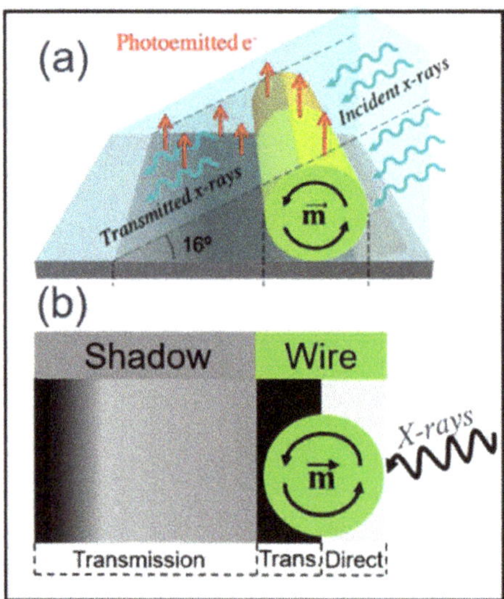

Figure 4. Schematic illustrations of (**a**) the principle of the dual sensitivity of Photoemission Electron Microscopy (PEEM) to detect direct photoemission and transmission data using X-ray Magnetic Circular Dichroism (XMCD) as a contrast mechanism, (**b**) magnetic contrast observed for direct photoemission and transmission. Adapted with permission from ref. [21]. Copyright 2016 Royal Society of Chemistry.

2.4. Micromagnetic Simulations

The understanding of 3D magnetic structures and their dynamics requires complementary micromagnetic modeling. In this work micromagnetic modeling of individual nanowires has been carried out using a finite-difference discretization scheme implemented in mumax3 software [33]. The material parameters and the crystal structure are detailed in Table 2.

Table 2. Materials parameters used in micromagnetic modeling: saturation magnetization, $\mu_0 M_s$, exchange stiffness, A_{ex}, exchange-correlation length, $l_{ex} = (2A_{ex}/\mu_0 M_s^2)^{1/2}$, crystalline symmetry, first magnetocrystalline anisotropy constant, K_1, and the direction of the magnetization easy axis with respect to the nanowire axis. Saturation magnetization and exchange stiffness for CoNi alloys have been obtained by linear interpolation with the Co content of the alloy [20].

Material	$\mu_0 M_s$ (T)	A_{ex} (pJ/m)	l_{ex} (nm)	Crystal Symmetry	K_1 (kJ m^{-3})	Magnetization Easy Axis (e.a.)
$Fe_{20}Ni_{80}$ [34]	1.0	10.8	5.2	-	0	-
Co(111) [34]	1.76	13.0	3.3	Cubic	−75	parallel to nanowire axis
Co(100) [34]	1.76	13.0	3.3	Uniaxial	450	e.a. at 75°–88° with nanowire axis
Co-hcp [34,35]	1.76	30.0	4.9	Uniaxial	450	e.a. at 75°–88° with nanowire axis
$Fe_{30}Co_{70}$ [36]	2.0	10.7	2.6	Cubic	10	polycrystalline textured, e.a. at 45° vs. nanowire axis *
Ni(111) [34]	0.61	3.4	4.8	Cubic	−4.8	parallel to nanowire axis
$Co_{85}Ni_{15}$ [20,35,37,38]	1.60	26.0	5.1	Uniaxial **	350	e.a. at 65–88° with nanowire axis
$Co_{65}Ni_{35}$ [20,35,37,38]	1.35	15.0	4.5	Uniaxial **	260	e.a. at 65° with nanowire axis
$Co_{35}Ni_{65}$ [20,35,37,38]	1.01	10.0	5.0	Cubic	2	parallel to nanowire axis

* In each grain, one easy axis (e.a.) is randomly set on the surface of a cone of 45 degrees with respect to the nanowire axis, and the second e.a. is placed with random orientation in the plane, normal to the first axis [39]. ** Polycrystalline structure was also used with no significant difference with the uniaxial anisotropy.

3. Magnetic Configurations of Cylindrical Nanowires with Large Shape Anisotropy

3.1. Magnetic Domain Configuration in Nanowires with Uniform Diameter

The magnetism of cylindrical nanowires is determined mainly by their shape (i.e., geometry) and magnetocrystalline anisotropy. A full understanding of the magnetization reversal process of individual nanowires and their arrays is essential to design and develop novel applications.

The soft magnetic nanowires, with *fcc* or *bcc* crystal symmetry, exhibit reduced crystalline anisotropy, lower than the axial shape anisotropy [40,41]. The elongated shape of the wire induces a natural axial anisotropy via the magnetostatic energy which favors a high stability of axial magnetic states [42]. Due to the strong magnetic shape anisotropy, these nanostructures present high coercivity and remanence when magnetized along their long axis.

The magnetic state of nanowires with cubic structures, i.e., Ni, Fe, and Py, is characterized by a predominant magnetization component along the nanowire axis with two open vortices at the nanowire ends (Figure 5a) which minimize the magnetostatic energy. The magnetization reversal takes place by domain wall propagation. The type of domain walls by which the nanowire demagnetizes is determined by the nanowire geometry (diameter) and material. While the thin nanowires demagnetize by transverse domain wall, the large-diameter nanowires demagnetize via Bloch point (previously called vortex) domain wall [29,42–44].

The incorporation of new elements into the system modifies both the structure and magnetic response of nanowires. One of the relevant materials is a Co-based alloy which presents different magnetic configurations as a function of the crystallographic structure, highly influenced by the preparation parameters (electrolyte, temperature, deposition, pH) [45]. In the case of NiCo wires, the magnetic properties are tuned through the composition of the alloy [38,46,47]. For less than 50% Co, the shape anisotropy predominates, allowing for a variation of coercivity as a function of composition, while for the content of Co > 50% the magnetocrystalline anisotropy becomes predominant and the magnetic properties are mainly determined by the crystallographic phases. A promising nanowire alloy is FeCo due to its high saturation magnetization and elevated Curie temperature, magnitudes that make it relevant in most technological applications, and specifically for a novel family of permanent magnets [36,48].

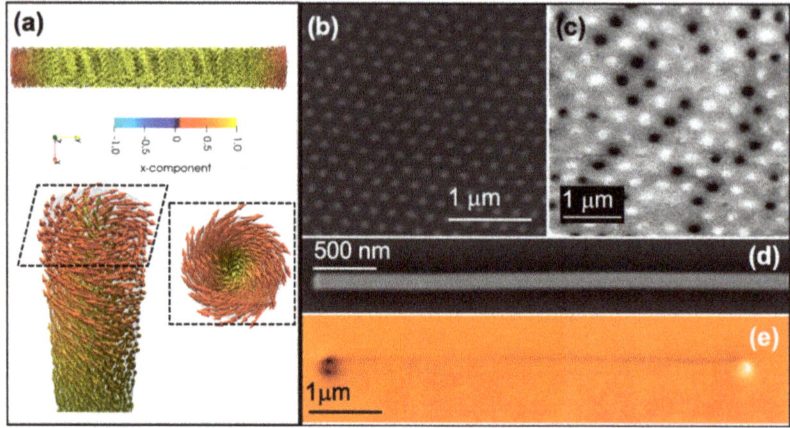

Figure 5. (**a**) Micromagnetic simulation of an individual FeCo nanowire presenting a longitudinal magnetization with vortices at the ends (top panel). Close-up images of the vortex structure are presented in the bottom panel, (**b**) SEM top view image of alumina template filled by magnetic nanowires, (**c**) Magnetic Force Microscopy (MFM) image of Ni nanowires embedded into the template (Reprinted with permission from ref. [49]. Copyright 2007 American Physical Society), (**d**) SEM image of a single CoNi nanowire, (**e**) MFM image of $Co_{35}Ni_{65}$ nanowire.

The first experimental investigations on shape anisotropy-dominated nanowires were done by MFM on Ni nanowires with *fcc* structure, 180 nm in diameter, embedded in alumina templates. The measurements were done at the top of the alumina template filled with magnetic nanowires. To remove the roughness at the top of the alumina substrate which can influence the MFM measurements the samples are mechanically polished. An example of a mechanically polished alumina template with the ends of the nanowires reaching the surface is presented in Figure 5b.

The MFM image in Figure 5c was taken after the sample and the tip were saturated in a negative magnetic field (black contrast). The white/black contrast corresponds to nanowires with the magnetization oriented up or down, respectively. When the magnetization of the nanowires points parallel/antiparallel to the tip field direction, we obtain black and white contrast, respectively. A simple magnetic phenomenological model allowed determining the magnetostatic interactions which strongly influence the remanence of the array [49].

The single-domain state in an individual nanowire (Figure 5d) with low magnetocrystalline anisotropy is presented in Figure 5e. The MFM image presents an individual $Ni_{65}Co_{35}$ nanowire with 120 nm in diameter where a uniform contrast is observed along the nanowire length, indicating a single longitudinal domain state. The bright/dark contrast at the ends of the nanowire is due to the presence of magnetostatic charges.

3.2. Magnetic Configurations in Nanowires with Tailored Geometry

In order to be used in 3D nanotechnological applications, e.g., data storage, sensing magnetomechanical actuation, or bio applications, the control of domain wall dynamics, nucleation, mobility, and pinning is paramount [1,50,51]. To pin the domain walls at certain positions along the nanowire length, several strategies have been considered. The approach consists of the creation of potential wells and barriers where the domain wall gets pinned. To do so, the nanowire geometry is altered creating constrictions along the length, artificial notches, anti-notches, defects, or diameter modulations that act as pinning sites for the domain walls.

Here, we discuss two types of geometrical constrictions: anti-notches (bamboo-type nanowires) (Figure 3b) and modulations in diameter (alternating diameters) (Figures 2b and 3

The considered modulated nanowires are made of FeCo alloys with high saturation magnetization (~1.8–2 T).

The first investigations were done by MFM on modulated $Fe_{30}Co_{65}Cu_5$ nanowires with small diameters deposited in modulated pores prepared by pulsed anodization in sulfuric acid (Figure 2a$_2$,b).

Figure 6 presents the morphology, the magnetic configuration, and micromagnetic simulations of an isolated FeCoCu modulated nanowire. The High-Resolution Transmission Electron Microscopy (HRTEM) images in (a) present the morphology of an isolated wire formed by alternating segments of few hundreds of nanometers with two distinct diameters 22 and 35 nm. The HRTEM data also revealed the crystallographic structure of the two types of segments. The nanostructures present a cubic (*bcc*) crystalline structure in both segments, but with higher texture along the (110) direction in the segments with a larger diameter.

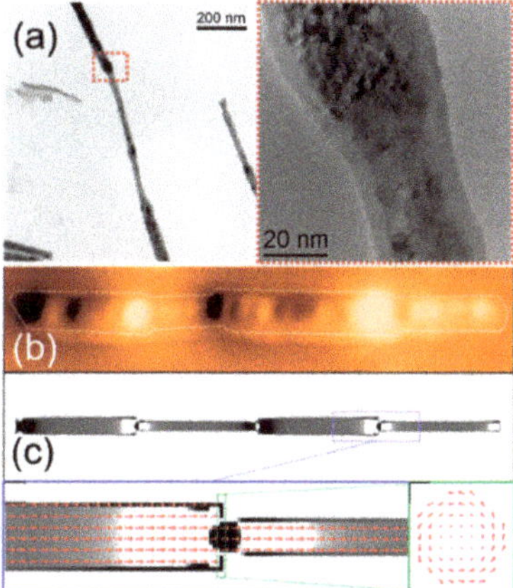

Figure 6. (**a**)-(left), HRTEM images of a modulated nanowire, (right), a close-up image of the area highlighted with a red square in (**a**)-(left). (**b**) MFM image of an isolated FeCoCu modulated nanowire (**c**) Simulated configuration at remanence for a modulated wire with two thick and two thin parts, equivalent to the experimental case. Adapted with permission from ref. [24]. Copyright 2015 IOP Publishing Ltd.

The magnetic configuration determined by MFM (Figure 6b) presents a single domain state that gives rise to strong contrast at the ends of the wire as well as at the transition region between segments of different diameters. The experimental data are supported by micromagnetic simulations (Figure 6c) where apart from the overall longitudinal configuration of magnetization, a curling effect is obtained at the transition regions between segments of different diameters. Moreover, the weaker bright/dark contrast, observed along segments, is correlated with higher roughness observed in modulated nanowires with smaller diameters [24].

A more detailed and defined geometry has been obtained for the nanowires with diameters over 100 nm (Figure 3) fabricated in the modulated pores obtained by pulsed anodization in oxalic acid (Figure 1b). The magnetization reversal of individual FeCoCu nanowires with diameters over 100 nm and the influence of tailored periodical geometrical modulations have been studied by the Magneto-Optical Kerr Effect (MOKE) [19].

Figure 7 shows the hysteresis loops for homogeneous diameter (a) and modulated in geometry (b and c) nanowires measured with the laser spot focused in the center of the nanowire. For the nanowire with a uniform diameter (a), a square hysteresis loop is observed with a sharp transition between two stable magnetic states at remanence through a single giant Barkhausen jump, suggesting the existence of a single domain structure with axial magnetization [38,52]. For the bamboo-type nanowires in (b), two abrupt and symmetric magnetization jumps are observed in each branch of the hysteresis loop. In the case of the modulated nanowire (c), the hysteresis loop shows the existence of a main Barkhausen jump together with several additional ones of smaller amplitude. The presence of several magnetization jumps in the modulated nanowires suggests the existence of metastable magnetic states during the magnetization reversal that could be correlated to their particular diameter modulation [19,52]. The MOKE measurements done at different spots along the nanowire's length and their angular dependence indicate that the demagnetization process takes place by the propagation of a single vortex domain wall which eventually is pinned at given modulations with a slightly higher energy barrier [19].

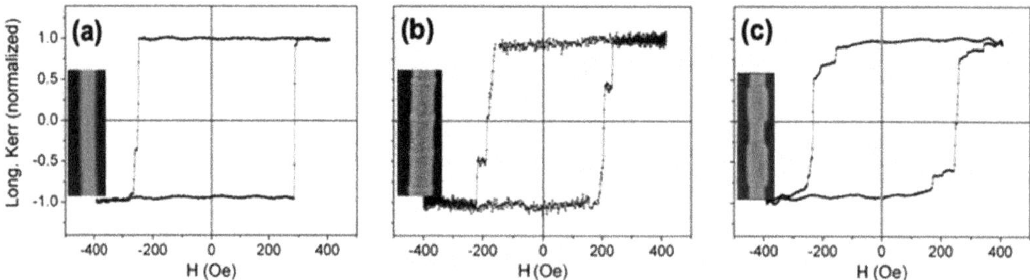

Figure 7. Magneto-Optical Kerr Effect (MOKE) hysteresis loops for uniform diameter (**a**) bamboo-type (**b**) and modulated diameter (**c**) nanowires. The insets show the SEM images of individual nanowires. Adapted with permission from ref. [19]. Copyright 2015 IOP Publishing Ltd.

A clear picture of the magnetic configurations in these two types of modulated in geometry nanowires was obtained by employing advanced microscopy techniques like Magnetic Force Microscopy (MFM), X-ray Magnetic Circular Dichroism combined with Photoemission Electron Microscopy (XMCD-PEEM), and Electron Holography. XMCD-PEEM offers the possibility in the cylindrical geometry of nanowires to obtain a full mapping of magnetic configuration. Due to the partial transmission of the X-ray beam through the wire, the magnetic state of the nanowire core is mapped onto the substrate providing simultaneous information of the magnetization distribution at the surface (direct photoemission from the wire) and inside the nanowires (photoemission from the substrate) (Figure 4a).

In addition, electron holography (EH) supplies information on the magnetic flux distribution of the internal magnetic structure and the stray fields outside the nanowire. The technique provides a quantitative analysis of both the crystallographic structure and the magnetic properties obtained on the same area, and at the nanoscale [22,53,54].

Figure 8a,b present the morphology and XMCD-PEEM measurements of a bamboo-type FeCoCu nanowire with anti-notches placed at about 400 nm along the nanowire axis. The X-ray diffraction data revealed that the FeCoCu wires present a *bcc* polycrystalline structure. The XMCD image in Figure 8b is characterized by a modulated profile along the entire length. The intensity profile in the surface region contains bright/dark local contrast matching the position of each modulation along the length. Between those local regions the surface shows a reduced grey contrast, characteristic of a longitudinal magnetization orientation perpendicular to the X-ray propagation vector. The enhanced magnetic contrast can be observed at the end of the wire and at given modulations, as shown in the insets of Figure 8 (i), (ii), and (iii) suggesting vortex-like domain walls pinned at the anti-notches. In the shadow, we get information about magnetization orientation inside the nanowire. A uniform grey contrast is observed in the main region of the shadow along the whole length of the nanowire. The vanishing grey contrast of the main region of the shadow reveals the longitudinal orientation of the magnetization inside the nanowire [21,55].

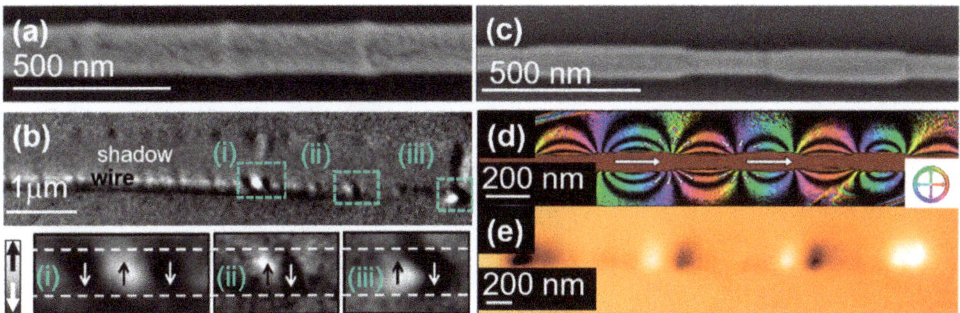

Figure 8. (**a**) SEM image of a bamboo FeCoCu nanowire, (**b**) XMCD-PEEM image of bamboo-type FeCoCu nanowire (**a**) oriented perpendicular to the incident X-rays (Adapted with permission from ref. [21]. Copyright 2016 Royal Society of Chemistry). The insets show a closer view at the local magnetic configuration marked in (**b**) by green dashed squares, (**c**) SEM image of a modulated FeCoCu nanowire, (**d**) magnetic flux images of modulated FeCoCu nanowire (**c**) reconstructed from the magnetic phase shift images, (**e**) MFM image of a modulated FeCoCu nanowire. Adapted with permission from ref. [22]. Copyright 2016, American Chemical Society.

A quantitative magnetic characterization has been performed on individual diameter-modulated FeCoCu nanowires with alternating segments of 100 and 140 nm (Figure 8c) by electron holography and MFM [22]. The analysis shows that the diameter-modulated geometry of the wires induces the formation of vortex-like structures and magnetostatic charges at the border between segments with different diameters, modifying the axial alignment of the magnetization in large-diameter segments. Furthermore, the magnetostatic charges influence the stray field distribution, inducing a flux-closure stray field configuration around large-diameter segments and keeping the demagnetizing field parallel to the nanowire's magnetization around small diameter segments (Figure 8d). The holography data complements the MFM data unveiling the origin of bright and dark contrast observed along the nanowire (Figure 8e).

A similar study was performed on two sets of Ni tri-segmented modulated nanowires, with two different diameters of $D_1 = 160$ nm (145 nm) and $D_2 = 130$ nm (100 nm) and lengths of $L_1 = 9.7$ µm (6 µm) and $L_2 = 1.6$ µm (5.5 µm) [56]. The experimental and modeling data revealed that the vortex domain wall nucleates and propagates along the nanowires. In the case of the nanowire formed by a thin segment encapsulated between two larger segments, the domain wall is trapped at the junction between the large and narrow diameter.

By employing a combination of Atomic Layer Deposition (ALD) and anodization, Prida et al. [57,58] synthesized alumina templates with hexagonally ordered pores having one well-defined geometrical modulation in the diameter (Figure 9a). Ni and $Fe_{50}Co_{50}$ nanowires were grown into the tailored alumina membranes by electrodeposition to replicate the geometry of the alumina templates with bi-segmented pores (Figure 9b). The nanowires with two distinct diameters (30 and 80 nm) show a polycrystalline crystallographic cubic structure, *fcc* (220) for Ni nanowires and *bcc* (110) for FeCo nanowires, respectively (Figure 9b-inset).

A uniaxial magnetization easy axis was determined for the bi-segmented nanowires with the magnetization reversing through the propagation, in steps, of a domain wall due to the geometrical modulation. This can be seen in the hysteresis loops and in the First-Order Reversal Curve (FORC) data (Figure 9c,d) of Ni nanowire arrays. The FORC diagram obtained for bi-segmented Ni nanowires (Figure 9c) shows a FORC distribution shape associated with a magnetic nanowire array with predominant shape anisotropy. The diagram is characterized by a single branch, which spreads widely parallel to the interaction field axis. However, due to the sharp modulation, this branch splits in two at different values of coercive field (see the red markings in Figure 9c). A FORC diagram for nanowires with a homogeneous diameter (80 nm) was also obtained (Figure 9d) showing

only one branch parallel to the interaction field axis. The magnetic experiments show th
two-step magnetization reversal process, ascribed to the segments of different diameter
in both Ni and FeCo nanowire arrays [57,59].

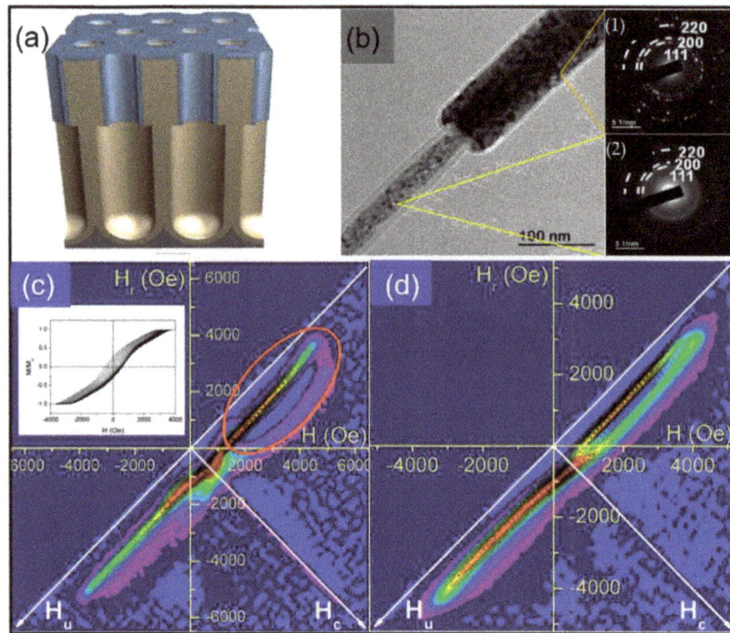

Figure 9. (**a**) Schematic view of highly ordered alumina template with modulated pores, (**b**) HRTE
image of a bi-segmented Ni nanowire. The insets show the Selected Area Electron Diffraction (SAE
patterns corresponding to the wide and narrow segments of the nanowire shown in (**b,c**) First-Ord
Reversal Curve (FORC) distribution diagram for the bi-segmented Ni nanowire arrays. The ins
shows the hysteresis curves used for calculating the respective FORC diagram (**d**). FORC distributi
diagram for the Ni nanowire arrays with 80 nm in diameter (Adapted with permission from ref. [5
Copyright 2019 Springer Nature.).

3.3. Micromagnetic Simulations of Nanowires with Tailored Geometry

Micromagnetic simulations are the necessary tool to assist the design of the geomet
and material properties of nanowires with the aim to induce the pinning/unpinnir
processes of domain walls and to understand the mechanisms behind them. In this sectic
we discuss several possibilities of geometrical constrictions in FeCo cylindrical nanowir
based on anti-notches (Figure 10a) and modulations in diameter (alternating segments wi
different diameters) in Figure 10b. The micromagnetic parameters are listed in Table 2.

The geometry and the remanent state of an individual FeCo nanowire with an
notches (bamboo-type) are shown in Figure 10a. Overall, the magnetization adopts a
axial orientation (except for the regions close to the anti-notches) determined main
by the shape anisotropy of the nanowire which is orders of magnitude larger than th
magnetocrystalline anisotropy of FeCo. In addition, this magnetic configuration shov
open vortices at each end of the nanowire. Along the nanowire surface, the magnetizatic
shows a smooth curling that reduces the formation of magnetostatic charges at the positior
of each anti-notch (marked by green arrows). This deviation of the magnetization from i
axial orientation agrees with the periodic contrast reported by MFM [55] and XMCD [2
data for FeCo-based bamboo nanowires.

Figure 10b presents the magnetization configuration, prior switching, at the surfa
of a single modulated polycrystalline FeCo nanowire with alternating segments of larg

diameter, 130 nm, and minor diameter (d = 40–100 nm) [39]. The demagnetization process in this nanowire begins in the segments of the larger diameter by means of the nucleation of the open vortex structures with arbitrary vorticity at the constrictions. These structures unpin from the constriction and propagate first inside the wide diameter segments and later inside the small diameter segment. When the difference between diameters is small the propagation in both segments, although consequent and not simultaneous, takes place at the same field value ("weak pinning").

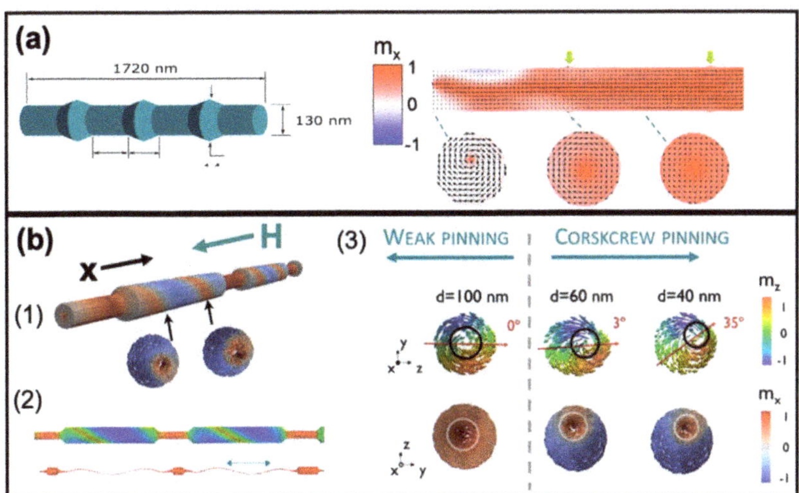

Figure 10. (**a**) Geometry of a FeCo nanowire with anti-notches (bamboo-type nanowire) on the left side and, a snapshot of one end of the nanowire at the remanent state on the right (Adapted with permission from ref. [21]. Copyright 2016 Royal Society of Chemistry.). The green arrows indicate the position of the anti-notches. The dashed lines show the location of the cross-sections. (**b**) (1) The magnetization at the surface of a modulated nanowire with segments of 130 nm in diameter and a smaller diameter d in a magnetic state before magnetization switching. The cross-section taken at the positions marked by arrows shows the skyrmion tube state. (2) The core of the skyrmion describes a helicoidal curve inside the segments of larger diameter. The segments of minor diameter remain uniformly magnetized. (3) magnetization configurations in the cross-sections of the segments with larger diameters in nanowires with segments of smaller diameters. Adapted with permission from ref. [39]. Copyright 2018 Royal Society of Chemistry.

For a larger difference in diameters between the two segments, i.e., strong pinning of magnetic vortex structures nucleated at constrictions, the first stage of the demagnetization process consists only in the propagation of the structures inside the segments of larger diameter. At a second stage (at more negative fields), the vortex structures de-pin from the constrictions and propagate inside segments of small diameter, i.e., the magnetization switches its axial orientation.

Importantly, the open vortex structures at the constrictions between large and small segments are formed with arbitrary chirality. As the magnetic field is increased further, the magnetic moments in the outer nanowire shell rotate towards field direction forming a spiral skyrmion tube (see Figure 10b-(1)). This also can be seen from the magnetization configurations in large segment cross-sections in Figure 10b-(1), the magnetization structure is formed by a magnetic skyrmion with the core pointing against the direction of the magnetic field and the shell pointing parallel to it. Once the demagnetization is completed in the large-diameter segments, the skyrmion structure remains pinned at the constrictions. Importantly, the skyrmion center along the segment is not located in the center of the nanowires but describes a spiral (see Figure 10b-(2)), a corkscrew magnetization pattern [39]. The formation of spiral for vortex/skyrmion tube center is a consequenc of the minimization of magnetostatic charges (magnetic poles) at the constriction. Thes charges are typically created at the constrictions and are redistributed along the nanowir length to minimize the energy. Arrot et al. [60] argue that these charges are proper t all magnetic nanowires but they should be particularly visible in nanowires with larg saturation magnetization where the minimization of magnetostatic energy is favorabl Similar topologically protected structures and the core-screw surface structure have bee recently reported in FeNi nanowires with chemical barriers and microwires [61–63].

3.4. Magnetic Configuration of Individual Nanowires with Chemical Notches

A controlled magnetization reversal can be achieved apart from geometrical constri tions, by chemical notches placed along the nanowire's length (Figure 11a).

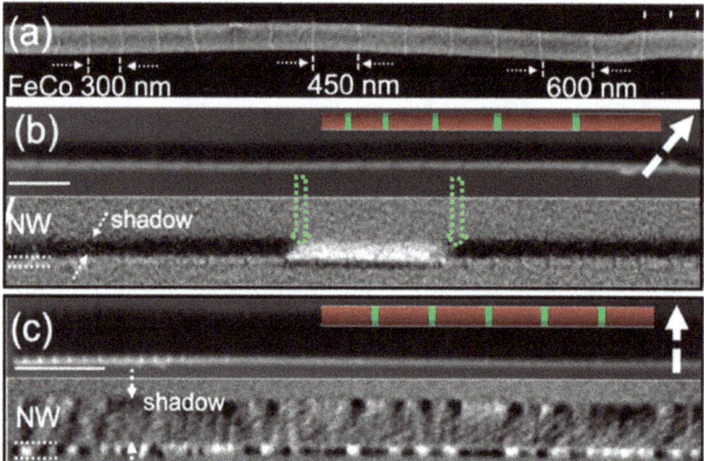

Figure 11. (**a**) SEM image of a multisegmented FeCo/Cu nanowire. Sequences of X-ray Absorptio Spectroscopy (XAS) (above, with an indication of the beam direction: dashed white arrows) an PEEM images of (**b**) multisegmented FeCo/Cu nanowire with a variable length of FeCo segment and (**c**) multisegmented FeCo/Cu nanowire with constant length of FeCo segments. The scale bar : 1 μm.

The engineering of the magnetic properties can be done not only through the archite ture of the nanostructure (combination of materials-layers) but also by controlling thei crystallography or composition and interfaces. Their fabrication is usually achieved by tw methods: a sequential deposition where two electrolytic baths are involved or by usin a single electrolytic bath and varying the electrodeposition potential or current densit to obtain a different composition in each layer [64]. The first method allows full contrc over the composition of individual layers/segments, while the second one produces bette interfaces between additional layers.

FeCo(x)/Cu (30 nm) nanowires were grown by electrodeposition in a three-electrod cell, at room temperature, from a single electrolytic bath (electrolyte (1) in Table 1). Th pH value of the electrolyte was maintained at about 3.0. The applied potential, versu the Ag/AgCl reference electrode, was alternately pulsed between −0.6V to deposit C and −1.8 V for different time periods to deposit the FeCo layers [65,66]. To be measure individually, the nanowires were released by chemical etching from the template.

Figure 11 displays the magnetic configuration, measured by XMCD-PEEM of tw individual FeCo/Cu multisegmented nanowires. Above each stack of XMCD-PEEN images, we show the direct X-ray Absorption Spectroscopy (XAS) image at the Co L absorption edge for chemical identification of the Cu segments. The XMCD-PEEM image present contrast both in the wire (dotted lines labeled NW) and in the shadow, due to the photoemission from the substrate after transmission through the wire core. Figure 11a,b shows the SEM image (a) and chemical (top panel) and magnetic contrast (bottom panel) (b) of a FeCo/Cu multisegmented nanowire with 120 nm in diameter, fixed Cu layers of 30 nm, and increasing length (from left to right) of FeCo segments (~200–1000 nm). The XMCD contrast along the wire suggests three longitudinal magnetic domains with magnetic moments pointing along the nanowire axis and separated by two domain walls pinned at the Cu constrictions as evidenced by green dotted arrows. Furthermore, the homogeneous contrast of opposite orientations between the nanowire and shadow observed in all the individual segments indicates a uniform axial magnetization at the nanowire surface and core.

A different magnetic configuration is presented in Figure 11c where, although the nanowire composition is similar to that presented in (a), the magnetic landscape is different. The images presented in Figure 11c present the chemical and magnetic contrast of a FeCo/Cu nanowire with constant FeCo and Cu segments of 250 nm and 50 nm, respectively, and 165 nm in diameter.

The top panel in Figure 11c shows the XAS image at the Co L_3 absorption edge for chemical identification of the different segments (modulated contrast on the wire and in the shadow). The magnetic contrast profile (Figure 11c-bottom panel) presenting bright/dark on the wire or dark/bright in the shadow matches the position of magnetic layers, suggesting the presence of a single vortex structure in each FeCo layer. The measurements indicate that the magnetic moments are pointing parallel (white contrast on the wire/dark in the shadow) or antiparallel (dark/white contrast in the shadow) to the polarization vector.

The examples presented above offer a comprehensive picture of the role played by the architecture of the nanostructure: on one hand, we have strongly coupled FeCo segments across the thin Cu layers which orient the magnetization along the long axis of the nanowire (Figure 11b) while in the second example, the reduced magnetostatic coupling due to the lower shape anisotropy determined by the short lengths of the segments and larger Cu spacer, makes each segment behaving as a single nanodot with the magnetization in a single vortex state (Figure 11c).

3.5. Manipulation of Magnetization Reversal by Magnetic and Electric Fields

The full control over the domains and domain walls in nanowires with circular cross-sections requires the determination of the conditions for their efficient and minimal manipulation by magnetic fields and spin-polarized currents.

The magnetization reversal following a ratchet effect was observed in multisegmented nanowires with increasing segment lengths presented in Figure 12 [65]. The unidirectional propagation, irrespective of the longitudinal field direction, is experimentally observed and confirmed by micromagnetic simulations in $Fe_{35}Co_{65}$/Cu multisegmented nanowires with fixed Cu layers of about 30 nm and variable lengths of FeCo segments (200–1000 nm). The magnetization reversal has been observed to proceed in few irreversible jumps at which magnetization reverses (as observed by MFM sensitive to surface mostly and detected by MOKE) as well confirmed in the whole nanowire segments as seen in XMCD-PEEM measurements (Figure 12a,b). The reversal process is always unidirectional, irrespective of the external field direction, initiating at the end of segments with a shorter length. Such ratchet effect originates in the broken symmetry induced by the shape anisotropy of increasing length of the FeCo segments and, like in a domino effect, it is promoted by the magnetostatic coupling between adjacent segments.

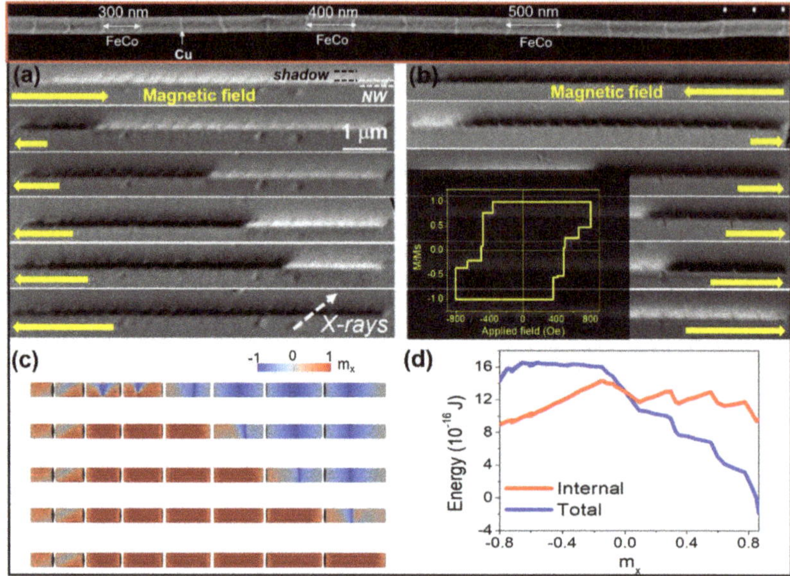

Figure 12. (**a,b**) Selected PEEM images under increasing applied field along the leftward (**a**) and rightward (**b**) polarity. The inset in (**b**) shows the reconstructed hysteresis loop. (**c**) Simulated magnetization configurations showing the sequential reversal during the reversal process. (**d**) Total and internal magnetic energies in FeCo/Cu nanowire as a function of the longitudinal magnetization component evaluated by the micromagnetic simulations during the ascending branch of the hysteresis loop. Adapted with permission from ref. [65]. Copyright 2018 American Chemical Society.

The micromagnetic simulations (Figure 12c,d) reveal a complex process where, although the switching is sequential from one segment to another, the magnetization process inside each segment takes place by the formation of vortices and skyrmion tubes followed by the final collapse of the internal core. The formation of skyrmion tubes with opposite chirality and strong topological protection may constitute the origin of pinned magnetic states. The plotted energy in Figure 12d displays the ratchet-like potential created by the increasing shape anisotropy, exchange energy, and pinning sites [65].

Reports on domain wall motion and control by means of electric current and thermo-magnetic switching in cylindrical nanowires are scarce and their analysis is limited [12,6...

Here, we introduce the results of our micromagnetic modeling on the manipulation and control of the polarity and vorticity of magnetic vortex structures (originating from their precursors at the nanowire ends at remanence) in a $Fe_{20}Ni_{80}$ cylindrical nanowire with a diameter of 100 nm by the simultaneous application of external fields and spin-polarized currents (Figure 13a). The remanent state is a uniform axial magnetic domain with open curled structures (Figure 5a, precursors of vortex domain wall or magnetic domains) with opposite chirality at each end of the nanowire (Figure 13c with J = 0). In all cases, vortices consist of an axially magnetized core (its direction defines the polarity of the vortex) and curling around the core (whose sense of rotation determines the vortex vorticity) as in the example in Figure 13b. The product of integer numbers (polarity by vorticity) is known chirality. See [68] for modeling and further details.

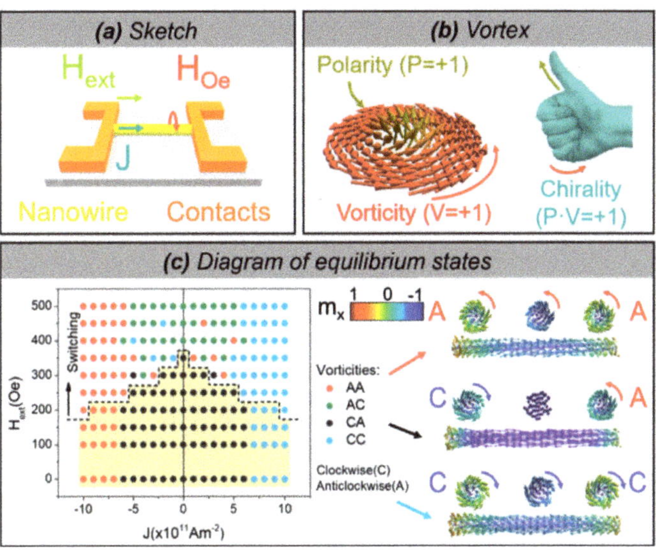

Figure 13. (**a**) Schematic illustration of a contacted nanowire in an experiment on magnetization dynamics. A spin-polarized current J flowing through the nanowire generates an Oersted field H_{oe} which adds to the external field H_{ext}. (**b**) The polarity and vorticity of a vortex are respectively defined by the directions of the magnetization in the inner core, and the sense of the rotation around this core (anticlockwise (A) or clockwise (C)). Their product determines the chirality which is equivalently determined by right/left-hand rule as right/left-handed vortex or anticlockwise/clockwise vortex. (**c**) Left side—a diagram of vortex stationary states as a function of the applied field and current. C/A stands for clockwise/anticlockwise vorticity. The threshold for the switching field of the axial component is indicated by the dashed line. Below this line (yellow-shaded region) no magnetization switching occurs, i.e., the shaded and non-shaded regions correspond to vortices with polarity +1 and −1 respectively. On the right side, three representative configurations of the magnetization for $H_{ex} = 0$ and vortex states AA, CA, and CC. Adapted with permission from ref. [68]. Copyright 2020 American Physical Society.

The diagram of stationary states is presented in Figure 13c and demonstrates the possibility for manipulation and control of the resulting vortex structures (i.e., both magnetization at the core and shell) with fields and currents. The axial magnetization component is only switched (at a critical value of the field) with the assistance of both the external field and current. For magnetic fields below the switching field, only the vorticities can be controlled by the current density and magnetic field. For the currents and fields for which the switching of the magnetization at the inner core occurs (above the dashed line), the sense of rotation (vorticity) of both vortex structures is regularly reversed from anticlockwise/clockwise (AC) to CA and vice versa for low currents. Since the Oersted field is not large enough to set the rotation sense in this case, they are determined by the direction of the magnetization and the resulting torque, and therefore the reversed pattern is found. In addition, there are some low currents for which CC and AA are found when the magnetization is switched. This ambiguity is not observed for higher current values, for which the chirality is fully determined by the Oersted field, either CC or AA. Conveniently, setting the current and field to zero values does not alter the vortex pattern and hence it is completely controlled. Notice that below the dashed line the chirality is determined by the vorticity, whereas above the dashed line the polarity is reversed, and the chirality is therefore determined by both the polarity and the vorticity. The switching of both polarity and vorticity, preserves the chirality of the initial remanent state, whereas the switching of only one of them leads to the chirality switching. Furthermore, the whole process takes

place in a few nanoseconds, consequently, excessive Joule heating could be prevented b
the use of short field and current pulses of this duration.

4. Magnetic Configurations of Cylindrical Nanowires with Large Magnetocrystalline Anisotropy

While in the case of nanowires with a cubic crystalline structure the shape anisotrop
and magnetostatic interactions mostly determine the overall magnetic response [34,69
in the case of Co-based nanowires with a hexagonal crystallographic structure the mag
netocrystalline anisotropy plays an important role as its easy axis can orient at differer
angles with respect to the nanowire axis. In Co nanowires, the crystalline structure ca
be tuned by varying the growth parameters such as applied voltage, pH, dimension
(diameter, length), or annealing and deposition under external magnetic fields. It has bee
shown that pH-controlled electroplating enables the switching between *fcc*- and *hcp*-C
phases, which modifies the magnetization easy axis from parallel to perpendicular to th
wires [45,70,71]. Furthermore, the magnetic properties of Co nanowires can also be tune
by adding different materials to the Co system. One of these materials, CoNi, allows th
design of the crystalline structure and magnetic configuration in the nanowire through th
composition of the alloy [20,38].

Figure 14a shows the XMCD image of a $Co_{85}Ni_{15}$ nanowire taken at a nearly perper
dicular configuration of the X-ray propagation vector with respect to the nanowire axi
The marked regions in the images labeled "wire" and "shadow" correspond respectivel
to photoemission from the nanowire surface and that from the substrate after transmissio
through the wire volume. The surface (wire, marked by white arrows in (a)) consists c
a sequence of segments with an azimuthal magnetic configuration of opposite contras
i.e., vortex-like structures with alternating chirality. The contrast observed in Figure 14a i
interpreted by the schematic illustrations in Figure 4. Due to the cylindrical geometry of th
nanowire and the grazing angle of the X-ray with respect to the sample surface (Figure 4
only a certain region of the nanowire surface (the bright contrast region) is directly expose
to the X-rays while the rest is only sensitive to the transmitted beam (the dark contra
region) [21]. This contrast, arising from the transmitted beam at the wire's surface ca
be seen in Figure 14a, marked by red arrows. The second type of contrast is found i
the shadow area due to the transmitted X-rays through the volume of the nanowire. Th
uniform grey contrast in the middle of the shadow (corresponding to the core of the wir
is an indication of the magnetization pointing perpendicular to the beam, i.e., along th
nanowire axis. Notice that for the same magnetic moment orientation, opposite contrast i
expected between direct and transmitted signals [20,21].

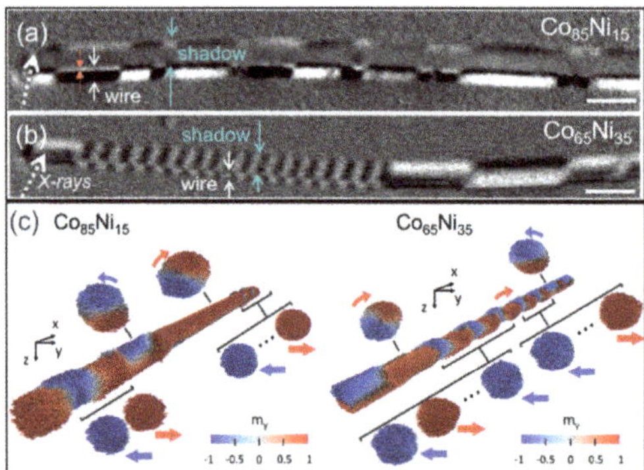

Figure 14. XMCD-PEEM images of (**a**) $Co_{85}Ni_{15}$ and (**b**) $Co_{65}Ni_{35}$ nanowires. The scale bar is 1 μr
and the arrows (dash white) indicate the incident X-ray beam, (**c**) micromagnetic simulations c
$Co_{85}Ni_{15}$ (left) and $Co_{65}Ni_{35}$ (right) nanowires. Adapted with permission from ref. [20]. Copyrigh
2017 American Physical Society.

A more complex magnetic configuration is found in $Co_{65}Ni_{35}$ nanowires (Figure 14b). The XMCD-PEEM image taken at Co L_3-edge shows two types of magnetic structures. On one hand, on the right side of the wire, a sequence of segments with opposite contrast similar to those in Figure 14a is observed. The increased contrast in the shadow indicates that the circular/vortex structures with ~1 μm in length, observed at the surface of the wire, penetrate into the volume, reducing the width/diameter of the longitudinal core. On the other hand, on the left side of the wire, a sequence of segments with shorter periodicity and alternating contrast (bright/dark at the surface, dark/bright in the core) is observed. The contrast in the shadow is opposite to that at the surface and remains constant in the transversal direction, showing that the magnetization state is homogenous along the circular cross-section of the wire. This is interpreted to correspond to alternating periodic transverse domains with the component of magnetization in the perpendicular direction to the nanowire axis. The width of the observed transverse domains is estimated to be about 150 nm and is very regular [20].

The coexistence of the metastable hybrid states in CoNi wires is confirmed by micromagnetic simulations which, furthermore, offers a better understanding of the magnetization processes. The simulations of $Co_{85}Ni_{15}$ nanowire are shown in Figure 14c (left side) where the equilibrium state is formed by a series of vortex and transverse domains along the nanowire. The magnetic landscape shows mainly vortex domains with opposite chirality and different lengths separated by shorter transverse domains with opposite directions close to the ends of the nanowire. For the $Co_{65}Ni_{35}$ nanowire (Figure 14c (right side)) a more complex domain structure is obtained. Again, a hybrid magnetic structure of transverse and vortex domains is observed along the nanowire. However, the transverse domains occupy a larger fractional volume, while the vortex domains mostly appear close to the ends.

Magnetic Nanowires with Modulated Anisotropy

The control and stabilization of domains and domain walls have been observed in multi-segmented nanowires [61,62,72,73] by alternating magnetic segments of different magnetic behavior or with non-magnetic metallic layers. These nanostructures can provide active channels for domain wall pinning or spin-wave manipulation. The multilayers formed by non-magnetic/magnetic layers can be used for domain wall pinning or confinement of specific domains [65]. The nanostructures formed by ferromagnetic/ferromagnetic segments offer the possibility to produce a system with alternating magnetocrystalline anisotropy which can pin domain walls between the segments or nucleate and stabilize them due to the influence of the neighboring material [72–74].

In this context, Co-based alloy nanowires are good candidates for tailoring designed magnetic anisotropies and consequently, magnetic behavior. The low magnetocrystalline anisotropy of cubic materials combined with the high anisotropy of the hexagonal symmetry of Co allows tailoring the magnetic properties of the system.

Multi-segmented CoNi/Ni nanowires with tailored alternating magnetic anisotropy have been fabricated and investigated with the aim to control the occurrence of different states by using the effect of confinement and interaction between segments. While in the case of Ni segments, due to the lack of magnetocrystalline anisotropy a single axial domain state is expected due to the predominant shape anisotropy, in the case of CoNi, due to the strong magnetocrystalline anisotropy energy constant of Co, the magnetic configurations can be tuned with respect to the composition (the strength of magnetocrystalline anisotropy and its easy axis) [73].

$Co_{85}Ni_{15}$/Ni wires with 140 nm in diameter, 1000 nm long Ni segments, and CoNi segments between 600 and 1400 nm in length were synthesized via electrochemical route. HRTEM data reveal that the Ni presents an *fcc* structure while the $Co_{85}Ni_{15}$ segments show an *hcp* crystalline structure oriented along the [010] direction. The magnetic configuration was imaged by XMCD-PEEM in the demagnetized state and at remanence after magnetizing the wires axially and perpendicularly.

Figure 15 presents the XMCD-PEEM images taken at the Co-edge, of CoNi/Ni multisegmented NWs with lengths of the CoNi segments of 600 nm (a), 1200 nm (b), and 1400 nm (c) after the previous demagnetization in a perpendicular magnetic field. In the shorter CoNi segments (600 nm) the image presents two types of contrasts: bright (dark in the shadow) or dark (bright in the shadow) suggesting single vortex states with different chiralities in different segments. In the CoNi with 1200 nm in length up to three vortices with alternating chirality (contrasts) are formed inside each segment (Figure 15b) while in the NWs with the longest CoNi segments (1400 nm), we observe the occurrence of either vortices or periodic transversal domains (Figure 15c). The inset of Figure 15c shows the XMCD profile over the CoNi segment marked by a dashed green square in (c). At the left side of the CoNi segment, first a vortex domain with a length of about 250 nm is formed (1) followed by the periodic transversal domains (2). The width of the transverse domains is estimated from the XMCD contrast profile across the wire (at the position of the green dashed line) to be about 105 nm (inset Figure 15c) [73].

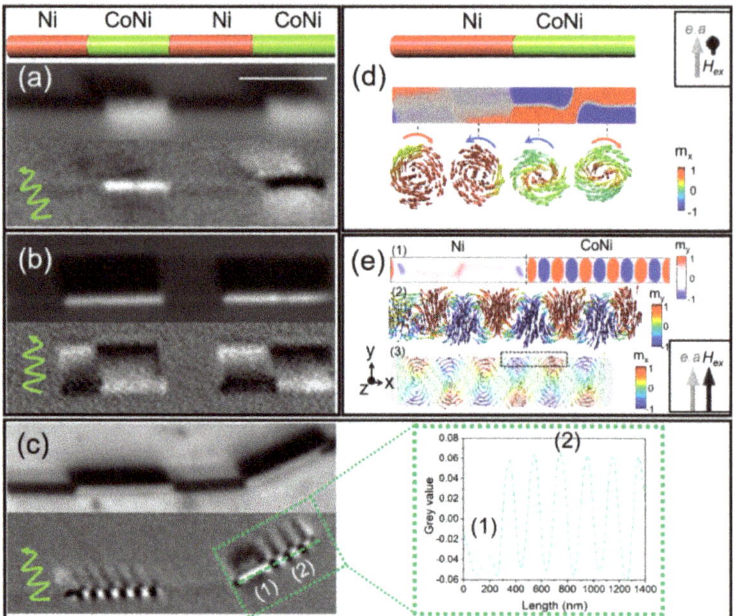

Figure 15. Chemical contrast (upper images) and XMCD-PEEM (lower images) contrast at Co L_3-edge of CoNi(x)/Ni(1000 nm) multisegmented NWs with (**a**) x = 600 nm, (**b**) x = 1200 nm and (**c**) x = 1400 nm. The inset in (**c**): XMCD profile of the CoNi segment at the right side in (c) presented at the position marked by the dashed horizontal green line. (**d**) Micromagnetic simulations of multi-segments with the applied magnetic field perpendicular to both, nanowire and magnetocrystalline easy axis (e.a.). (**e**) Micromagnetic simulations of multi-segments with the applied magnetic field parallel to the magnetocrystalline easy axis and perpendicular to the nanowire. Scale bar: 1 μm. Adapted with permission from ref. [73]. Copyright 2020 American Chemical Society.

The micromagnetic simulations (Figure 15d,e) reveal the origin of different magnetic configurations in the multisegmented nanowire. Although the Ni segment is a near single domain state, its magnetization presents a small magnetization curling at the surface. Its magnetic state is formed prior to the formation of magnetic structures in CoNi and determines the state of it. Interacting Ni segments (separated by short CoNi segments) prefer the same vortex chirality while the almost non-interacting ones (separated by long CoNi segments) prefer the formation of the alternating chirality. As seen in Figure 15d, the vortex chirality observed in CoNi domains mimics the chirality formed in the Ni segment in order to avoid magnetostatic charges at the transition. Furthermore, the remanent states of CoNi segments depend on the direction of the previously applied field. For the magnetic field applied perpendicular to both the nanowire and the anisotropy easy axis, the vortex domains expand inside the segment following the structures formed at the interfaces and result in one vortex domain for short segments and two or more vortex structures for longer segments. When the field is applied perpendicular to the nanowire but along the anisotropy easy axis, transverse domains are formed (Figure 15e-(1), (2)). They are separated by vortex domain structures with the core at the surface pointing perpendicular to it (Figure 15e-(3)) [73]. This configuration can also be considered as a chain of multi-vortex domains with the cores pointing perpendicular to the nanowire axis and transverse domain walls in between [54].

In a similar CoNi/Ni system, the strength of magnetocrystalline anisotropy of CoNi segments was lowered by reducing the Co content in the alloy. Figure 16 presents $Co_{65}Ni_{35}$ 2.1 µm long segments separated by Ni segments with 800 nm in length. The diameter of the multisegmented structure is 130 nm. The $Co_{65}Ni_{35}$ segment composition, the quality of interfaces, and the crystallographic structure were determined by HRTEM, confirming that Ni segments crystallize in *fcc* structure along (110), while CoNi alloy seems to grow epitaxially onto Ni segments [75].

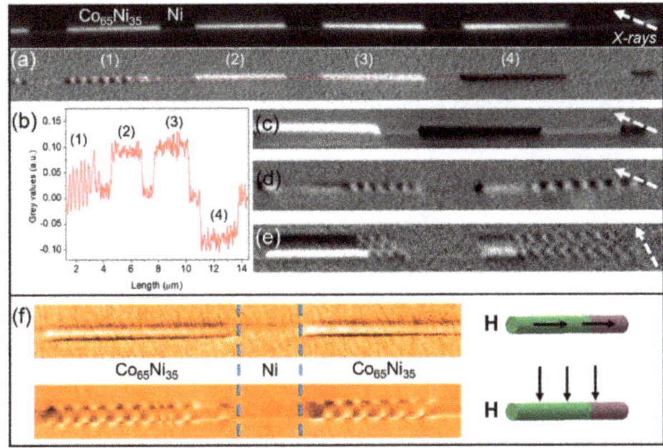

Figure 16. Chemical contrast of a multisegmented $Co_{65}Ni_{35}$/Ni nanowire (top panel). (**a**) XMCD-PEEM image of a $Co_{65}Ni_{35}$/Ni nanowire and (**b**) the contrast profile of CoNi segments in (**a**), labeled (1) to (4), with different magnetic structures. (**c–e**) XMCD-PEEM images of $Co_{65}Ni_{35}$/Ni nanowires showing different magnetic configurations. (**f**) MFM images taken in remanence (adapted from [75]), after applying a magnetic field of 1.8 T in the axial direction (top panel) and perpendicular to the nanowire (bottom panel).

Figure 16 shows the chemical contrast (top panel) and magnetic images ((a)–(f)) of $Co_{65}Ni_{35}$/Ni nanowires taken in remanence state. Figure 16a presents the XMCD-PEEM image of a CoNi/Ni nanowire measured at Co L_3-edge with the beam (marked by the dashed white arrow) at about 20 deg. with respect to the nanowire axis. The presented nanowire is formed by four main CoNi segments displaying two types of magnetic configurations visible at the surface of the nanowire. Due to the orientation of the beam to the nanowire axis, the XMCD is sensitive to the single longitudinal domain state as imaged in segments (2) to (4) and (c) with the magnetization oriented parallel (bright contrast) or antiparallel (dark contrast) to the X-rays (i.e., along the nanowire axis). In the other segments (1), we observed a periodic alternating contrast, similar to the transversal domains presented in Figures 14b and 15c. However, in this orientation of the X-ray propagation vector, almost parallel to the nanowire axis, only the magnetic moments oriented in the direction of the beam (along the wire axis) are observed whic[h] excludes the possibility of imaging transversal magnetic domains in the CoNi segment[s]. Figure 16b presents the contrast profile taken along the red dotted line in the nanowir[e] presented in (a): periodic magnetic structures with 140 nm in length (left side) followe[d] by longitudinal single domains with different orientations. Figure 16c,d show closed u[p] XMCD-PEEM images of different parts of the multisegmented nanowire presenting eithe[r] longitudinal single domain structure (c) or a mixture of vortex and periodical domain[s] (d). The same type of mixed magnetic configuration is presented in Figure 16e, this tim[e] measured with the beam oriented almost perpendicular to the nanowire axis. This specifi[c] orientation of the X-rays, almost perpendicular to the wire, gives us access to the "shadow[,"] which allows us a second set of information about the core of the measured wire. From th[e] alternate bright/dark contrast on the wire (dark/bright on the shadow) we conclude tha[t] the magnetization in CoNi segments consists of a series of periodic transversal domain[s] with about 90 nm in length, similar to those observed in $Co_{85}Ni_{15}$ segments. As reveale[d] by micromagnetic simulations (Figure 15e) the magnetization between transversal domain[s] consists of vortices with the core oriented perpendicular to the nanowire axis (the so-calle[d] "surface vortices"). This is in agreement with the XMCD data presented in Figure 16a-(1),[d] where the observed alternating dark/bright contrast at the surface of the wire is consisten[t] with the surface vortices predicted by simulations (Figure 15e-(3)), black dotted square[d] area) and observed experimentally by Electron Holography in CoNi [54]. By varying th[e] direction of the X-ray propagation vector with respect to the nanowire axis, XMCD-PEE[M] allows us to get information about both transversal magnetic domains (X-rays oriente[d] perpendicular to the wire, Figure 16e) and the domain walls in between, surface vortice[s] (X-rays oriented along the wire axis, Figure 16a,d).

The same magnetic configuration, i.e., single or periodic domains, was observed b[y] MFM [75]. Here, the magnetic configuration of $Co_{65}Ni_{35}$ could be tuned by magneti[c] fields (saturating or demagnetizing) applied parallel or perpendicular to the nanowir[e] axis [75]. Figure 16f presents the MFM pictures of the same multisegmented nanowire (tw[o] Ni segments and one CoNi segment), taken after the magnetic field was applied paralle[l] (Figure 16f-upper panel) or perpendicular (Figure 16f-bottom panel) to the nanowire axi[s]. Depending on the field history, the magnetization state of $Co_{65}Ni_{35}$ changes either in [a] single domain state (Figure 16f-upper panel) or periodic domains (Figure 16f-bottom pane[l]).

The multisegmented wires with modulated anisotropy presented in this section sho[w] a complex interplay of different energetic contributions and geometry determining th[e] resulting magnetic structures. This determines the different factors which should be take[n] into account for the design of magnetic nanowires with a certain magnetic configura[-] tion: geometry (diameter, length), the combined materials (interplay between shape an[d] magnetocrystalline anisotropy), and field history.

5. Discussion and Summary

Multiple cylindrical modulated magnetic nanowires with different materials an[d] geometries were prepared by electroplating filling the pores of anodic aluminum oxi[de] (AAO) membranes. The AAO membranes with modulated pores and various diamete[r] were obtained by pulsed anodization in sulfuric or oxalic aqueous solution. The sing[le] element and alloy nanowires, as well as multisegmented ones, were grown inside th[e] nanopores, at room temperature, by electrodeposition using different electrolytes.

The results show that in the case of nanowires with uniform diameter and sma[ll] magnetocrystalline anisotropy the magnetic configuration at remanence consists of a sing[le] domain with open vortices at the ends. These nanowires (Ni and Fe-based) are interestin[g] candidates to study the motion of domain walls.

Large magnetic anisotropy nanowires show a more complex behavior in terms o[f] magnetic configurations. Co-based nanowires with an hcp crystallographic structure ca[n] be prepared with magnetocrystalline anisotropy almost perpendicular to the nanowi[re] axis and typically present vortex domains with alternating chirality. The combination o[f] an almost perpendicular to nanowire anisotropy and the circular symmetry promotes the spontaneous development of vortex domains with magnetic moments following a circumferential path at the surface but staying longitudinal in the core. Furthermore, the use of different alloys with tailored compositions can alter the magnetic configuration going from axial domains to a combination of vortex and transverse domains, as shown in the example of CoNi alloy nanowires with different compositions. The control between different structures in some cases can be achieved by perpendicularly applied magnetic fields.

Multisegmented/multilayered nanowires offer an additional route to control the magnetization pattern via confinement and pinning. Particularly, further engineering of the nanowires can be provided by creating modulations/constrictions in diameter with the aim to achieve the pinning of domain walls at the constrictions. By altering the geometry of the nanowire, imprinting geometrical modulations (notches/anti-notches, alternating segments with different diameters) along the nanostructure's length the magnetic configuration changes. Although the inner magnetization stays unchanged (i.e., along the nanowire axis) surface chiral structures get pinned at the geometrical modulations distributed along the nanowire's length.

Both micromagnetic simulations and imaging revealed that in the case of modulated in geometry wires with high saturation magnetization, the magnetic configuration and pinning can be tailored further by playing with the diameter of the segments. Moreover, micromagnetic simulations show that when the diameter difference is large, we observe the formation of topologically protected structures, where the magnetization of the large-diameter segment forms a skyrmion tube with a core position in a helical modulation along the nanowire. This leads to an increase in the coercive field, as compared to the nanowires with uniform diameter, associated with the occurrence of a novel pinning type, i.e., "corkscrew" mechanism.

The nanowires also can be prepared with a well-controlled multilayer structure with alternating magnetic or magnetic/non-magnetic materials. The former allows the modulation of the magnetic anisotropy, while the latter can be used for domain wall pinning or confinement of different domains.

Engineering the nanowires with alternating magnetic/non-magnetic segments provides a way to control magnetization confinement and interactions by tailoring the lengths of the ferromagnetic segment and the thickness of the non-magnetic layer. Indeed, the magnetization structure in nanowires of high saturation and short segments can change from the longitudinal domains in a single material nanowire to vortex configurations confined in the segments. Moreover, the magnetization can be easily controlled by magnetic fields as shown in FeCo segments. If the length of FeCo segments, separated by 30 nm Cu layers, is gradually increased a ratchet system is formed where the magnetization reversal in neighboring segments propagates sequentially in steps starting from the shorter segments, irrespectively of the applied field direction. For FeCo segments with a constant length, separated by 50 nm Cu layers, the segments present a single vortex state with alternating chirality.

On other hand, the magnetic anisotropy was modulated in cylindrical nanowires by alternating two ferromagnetic materials with different magnetocrystalline anisotropy. This strategy shows a rich and complex behavior. The XMCD-PEEM imaging of multisegmented Ni/CoNi nanowires reveals that the magnetic structure in the Ni segment is in a dominantly axial magnetic state while in CoNi segments it depends on their length. Although in an almost single domain configuration, Ni state appeared to have an important impact on the CoNi, through magnetic interaction and magnetochiral effects, as revealed by micromagnetic simulations. The field history is revealed as another important factor, responsible for the appearance of different domain patterns (vortex domains or perpendicular/transversal domains separated by vortices on the surface) in large CoNi segments.

Future applications will require efficient manipulation of magnetic structures in individual nanowires. In this review, we presented many examples of how magnetic domain structure is changed by geometry, magnetocrystalline anisotropy, and applying field direc-

tion. At the same time, future energy-efficient applications will require the use of curre
which is scarcely investigated. Here we presented the results of micromagnetic simulatio
in a Py nanowire under the simultaneous action of the magnetic field and spin-polarize
currents. The results show that the current alone enlarges or reduces the open vort
structures formed at the ends of the nanowire depending on the rotational sense of th
associated Oersted field. Large current densities can set the vorticity along the who
nanowire in the desired direction. The switching of magnetization in the core is sole
achieved by the application of a simultaneous external field, lower than the coercive fie
of the nanowire.

In conclusion, our results demonstrate a complex interplay of different energet
contributions and geometry determining the resulting magnetic configurations. We stre
the importance of different factors which should be taken into account for the design
magnetic nanowires. Their tunability makes them excellent candidates for new applicatio
where full control over the magnetization reversal is mandatory.

Author Contributions: Conceptualization, C.B., investigation C.B., J.A.F.-R., R.P.d.R., A.A., O.C.-M.V.; resources, O.C.-F., M.V.; writing—original draft preparation, C.B., J.A.F.-R.; writing—revie and editing, C.B., J.A.F.-R., R.P.d.R., A.A., O.C.-F., M.V.; funding acquisition, O.C.-F., M.V. All autho have read and agreed to the published version of the manuscript.

Funding: This research was funded by the Regional Government of Madrid under the framework the project S2018/NMT-4321 NANOMAGCOST-CM and by the Spanish Ministry of Science and Inr vation through the projects MAT2016-76824-C3-1-R and PID2019-108075RB-C31/AEI/10.13039/501 0011033.

Acknowledgments: We acknowledge the service from the MiNa Laboratory at the Institute of Mic and Nanotechnology (IMN). The PEEM experiments were performed at the CIRCE beamline the ALBA Synchrotron with the collaboration of ALBA staff. For their contribution to the PEE experiments, we especially thank L. Aballe, M. Foerster and A. Fraile Rodriguez.

Conflicts of Interest: The authors declare no conflict of interest.

References

1. Streubel, R.; Fischer, P.; Kronast, F.; Kravchuk, V.P.; Sheka, D.D.; Gaididei, Y.; Schmidt, O.G.; Makarov, D. Magnetism in curve geometries. *J. Phys. D Appl. Phys.* **2016**, *49*, 363001. [CrossRef]
2. Braun, H.-B.; Charilaou, M.; Löffler, J.F. Skyrmion lines, monopoles, and emergent electromagnetism in nanowires. *Magnetic Nano- and Microwires*, 2nd ed.; Vázquez, M., Ed.; Elsevier: Amsterdam, The Netherlands, 2020; pp. 381–40 ISBN 978-0-08-102832-2.
3. Fischer, P.; Sanz-Hernández, D.; Streubel, R.; Fernández-Pacheco, A. Launching a new dimension with 3D magnetic nanostructur *APL Mater.* **2020**, *8*, 010701. [CrossRef]
4. Donnelly, C.; Scagnoli, V. Imaging three-dimensional magnetic systems with X-rays. *J. Phys. Condens. Matter* **2020**, *32*, 21300 [CrossRef]
5. da Câmara Santa Clara Gomes, T.; Marchal, N.; Abreu Araujo, F.; Piraux, L. Spin Caloritronics in 3D Interconnected Nanow Networks. *Nanomaterials* **2020**, *10*, 2092. [CrossRef]
6. Guzmán-Mínguez, J.C.; Ruiz-Gómez, S.; Vicente-Arche, L.M.; Granados-Miralles, C.; Fernández-González, C.; Mompeán, García-Hernández, M.; Erohkin, S.; Berkov, D.; Mishra, D.; et al. FeCo Nanowire–Strontium Ferrite Powder Composites f Permanent Magnets with High-Energy Products. *ACS Appl. Nano Mater.* **2020**, *3*, 9842–9851. [CrossRef]
7. Pierrot, A.; Béron, F.; Blon, T. FORC signatures and switching-field distributions of dipolar coupled nanowire-based hysterons *Appl. Phys.* **2020**, *128*, 093903. [CrossRef]
8. Zamani Kouhpanji, M.R.; Ghoreyshi, A.; Visscher, P.B.; Stadler, B.J.H. Facile decoding of quantitative signatures from magne nanowire arrays. *Sci. Rep.* **2020**, *10*, 15482. [CrossRef] [PubMed]
9. Alsharif, N.A.; Aleisa, F.A.; Liu, G.; Ooi, B.S.; Patel, N.; Ravasi, T.; Merzaban, J.S.; Kosel, J. Functionalization of Magne Nanowires for Active Targeting and Enhanced Cell-Killing Efficacy. *ACS Appl. Bio Mater.* **2020**, *3*, 4789–4797. [CrossRef]
10. Sharma, A.; Zhu, Y.; Thor, S.; Zhou, F.; Stadler, B.; Hubel, A. Magnetic Barcode Nanowires for Osteosarcoma Cell Contr Detection and Separation. *IEEE Trans. Magn.* **2013**, *49*, 453–456. [CrossRef]
11. Wolf, D.; Rodriguez, L.A.; Béché, A.; Javon, E.; Serrano, L.; Magen, C.; Gatel, C.; Lubk, A.; Lichte, H.; Bals, S.; et al. 3D Magne Induction Maps of Nanoscale Materials Revealed by Electron Holographic Tomography. *Chem. Mater.* **2015**, *27*, 6771–67 [CrossRef] [PubMed]

2. Schöbitz, M.; De Riz, A.; Martin, S.; Bochmann, S.; Thirion, C.; Vogel, J.; Foerster, M.; Aballe, L.; Menteş, T.O.; Locatelli, A.; et al. Fast Domain Wall Motion Governed by Topology and Œrsted Fields in Cylindrical Magnetic Nanowires. *Phys. Rev. Lett.* **2019**, *123*, 217201. [CrossRef]

3. Chu, S.-Z.; Wada, K.; Inoue, S.; Isogai, M.; Yasumori, A. Fabrication of Ideally Ordered Nanoporous Alumina Films and Integrated Alumina Nanotubule Arrays by High-Field Anodization. *Adv. Mater.* **2005**, *17*, 2115–2119. [CrossRef]

4. Lee, W.; Ji, R.; Gösele, U.; Nielsch, K. Fast fabrication of long-range ordered porous alumina membranes by hard anodization. *Nat. Mater.* **2006**, *5*, 741–747. [CrossRef]

5. Lee, W.; Kim, J.-C.; Gösele, U. Spontaneous Current Oscillations during Hard Anodization of Aluminum under Potentiostatic Conditions. *Adv. Funct. Mater.* **2010**, *20*, 21–27. [CrossRef]

6. Vega, V.; García, J.; Montero-Moreno, J.M.; Hernando, B.; Bachmann, J.; Prida, V.M.; Nielsch, K. Unveiling the Hard Anodization Regime of Aluminum: Insight into Nanopores Self-Organization and Growth Mechanism. *ACS Appl. Mater. Interfaces* **2015**, *7*, 28682–28692. [CrossRef] [PubMed]

7. Alraddadi, S.; Hines, W.; Yilmaz, T.; Gu, G.D.; Sinkovic, B. Structural phase diagram for ultra-thin epitaxial Fe_3O_4/MgO(0 0 1) films: Thickness and oxygen pressure dependence. *J. Phys. Condens. Matter* **2016**, *28*, 115402. [CrossRef]

8. Lee, W.; Park, S.-J.J. Porous Anodic Aluminum Oxide: Anodization and Templated Synthesis of Functional Nanostructures. *Chem. Rev.* **2014**, *114*, 7487–7556. [CrossRef]

9. Palmero, E.M.; Bran, C.; Del Real, R.P.; Vázquez, M. Vortex domain wall propagation in periodically modulated diameter FeCoCu nanowire as determined by the magneto-optical Kerr effect. *Nanotechnology* **2015**, *26*. [CrossRef]

10. Bran, C.; Fernandez-Roldan, J.A.; Palmero, E.M.; Berganza, E.; Guzman, J.; Del Real, R.P.; Asenjo, A.; Fraile Rodríguez, A.; Foerster, M.; Aballe, L.; et al. Direct observation of transverse and vortex metastable magnetic domains in cylindrical nanowires. *Phys. Rev. B* **2017**, *96*, 125415. [CrossRef]

11. Bran, C.; Berganza, E.; Palmero, E.M.; Fernandez-Roldan, J.A.; Del Real, R.P.; Aballe, L.; Foerster, M.; Asenjo, A.; Fraile Rodríguez, A.; Vazquez, M. Spin configuration of cylindrical bamboo-like magnetic nanowires. *J. Mater. Chem. C* **2016**, *4*, 978–984. [CrossRef]

12. Rodríguez, L.A.; Bran, C.; Reyes, D.; Berganza, E.; Vázquez, M.; Gatel, C.; Snoeck, E.; Asenjo, A. Quantitative Nanoscale Magnetic Study of Isolated Diameter-Modulated FeCoCu Nanowires. *ACS Nano* **2016**, *10*, 9669–9678. [CrossRef]

13. Minguez-Bacho, I.; Rodriguez-López, S.; Vázquez, M.; Hernández-Vélez, M.; Nielsch, K. Electrochemical synthesis and magnetic characterization of periodically modulated Co nanowires. *Nanotechnology* **2014**, *25*, 145301. [CrossRef]

14. Iglesias-Freire, Ó.; Bran, C.; Berganza, E.; Mínguez-Bacho, I.; Magén, C.; Vázquez, M.; Asenjo, A. Spin configuration in isolated FeCoCu nanowires modulated in diameter. *Nanotechnology* **2015**, *26*, 395702. [CrossRef] [PubMed]

15. Bran, C.; Palmero, E.M.; Li, Z.-A.; Del Real, R.P.; Spasova, M.; Farle, M.; Vázquez, M. Correlation between structure and magnetic properties in Co_xFe_{100-x} nanowires: The roles of composition and wire diameter. *J. Phys. D Appl. Phys.* **2015**, *48*. [CrossRef]

16. Moraes, S.; Navas, D.; Béron, F.; Proenca, M.; Pirota, K.; Sousa, C.; Araújo, J. The Role of Cu Length on the Magnetic Behaviour of Fe/Cu Multi-Segmented Nanowires. *Nanomaterials* **2018**, *8*, 490. [CrossRef]

17. García, J.; Vega, V.; Iglesias, L.; Prida, V.M.; Hernando, B.; Barriga-Castro, E.D.; Mendoza-Reséndez, R.; Luna, C.; Görlitz, D.; Nielsch, K. Template-assisted Co-Ni alloys and multisegmented nanowires with tuned magnetic anisotropy. *Phys. Status Solidi* **2014**, *211*, 1041–1047. [CrossRef]

18. Abreu Araujo, F.; Piraux, L. Spin-Transfer-Torque Driven Vortex Dynamics in Electrodeposited Nanowire Spin-Valves. *SPIN* **2017**, *7*, 1740007. [CrossRef]

19. Staňo, M.; Fruchart, O. Magnetic Nanowires and Nanotubes. *Handb. Magn. Mater.* **2018**, *27*, 155–267. [CrossRef]

20. Aballe, L.; Foerster, M.; Pellegrin, E.; Nicolas, J.; Ferrer, S. The ALBA spectroscopic LEEM-PEEM experimental station: Layout and performance. *J. Synchrotron Radiat.* **2015**, *22*, 745–752. [CrossRef]

21. Kimling, J.; Kronast, F.; Martens, S.; Böhnert, T.; Martens, M.; Herrero-Albillos, J.; Tati-Bismaths, L.; Merkt, U.; Nielsch, K.; Meier, G. Photoemission electron microscopy of three-dimensional magnetization configurations in core-shell nanostructures. *Phys. Rev. B Condens. Matter Mater. Phys.* **2011**, *84*. [CrossRef]

22. Scholl, A.; Ohldag, H.; Nolting, F.; Stöhr, J.; Padmore, H.A. X-ray photoemission electron microscopy, a tool for the investigation of complex magnetic structures (invited). *Rev. Sci. Instrum.* **2002**, *73*, 1362–1366. [CrossRef]

23. Vansteenkiste, A.; Van de Wiele, B. MuMax: A new high-performance micromagnetic simulation tool. *J. Magn. Magn. Mater.* **2011**, *323*, 2585–2591. [CrossRef]

24. Ivanov, Y.P.; Vázquez, M.; Chubykalo-Fesenko, O. Magnetic reversal modes in cylindrical nanowires. *J. Phys. D Appl. Phys.* **2013**, *46*, 485001. [CrossRef]

25. Moreno, R.; Evans, R.F.L.; Khmelevskyi, S.; Muñoz, M.C.; Chantrell, R.W.; Chubykalo-Fesenko, O. Temperature-dependent exchange stiffness and domain wall width in Co. *Phys. Rev. B* **2016**, *94*, 104433. [CrossRef]

26. Bran, C.; Ivanov, Y.P.; García, J.; Del Real, R.P.; Prida, V.M.; Chubykalo-Fesenko, O.; Vazquez, M. Tuning the magnetization reversal process of FeCoCu nanowire arrays by thermal annealing. *J. Appl. Phys.* **2013**, *114*, 043908. [CrossRef]

27. Kronmüller, H.; Fischer, R.; Hertel, R.; Leineweber, T. Micromagnetism and the microstructure in nanocrystalline materials. *J. Magn. Magn. Mater.* **1997**, *175*, 177–192. [CrossRef]

38. Vega, V.; Böhnert, T.; Martens, S.; Waleczek, M.; Montero-Moreno, J.M.; Görlitz, D.; Prida, V.M.; Nielsch, K. Tuning the magnetic anisotropy of Co–Ni nanowires: Comparison between single nanowires and nanowire arrays in hard-anodic aluminum oxide membranes. *Nanotechnology* **2012**, *23*, 465709. [CrossRef] [PubMed]

39. Fernandez-Roldan, J.A.; Perez del Real, R.; Bran, C.; Vazquez, M.; Chubykalo-Fesenko, O. Magnetization pinning in modulated nanowires: From topological protection to the "corkscrew" mechanism. *Nanoscale* **2018**, *10*, 5923–5927. [CrossRef] [PubMed]

40. Sellmyer, D.J.; Zheng, M.; Skomski, R. Magnetism of Fe, Co and Ni nanowires in self-assembled arrays. *J. Phys. Condens. Matter* **2001**, *13*, R433–R460. [CrossRef]

41. Fernandez-Roldan, J.A.; Ivanov, Y.P.; Chubykalo-Fesenko, O. Micromagnetic modeling of magnetic domain walls and domains in cylindrical nanowires. In *Magnetic Nano- and Microwires*; Woodhead Publishing: Amsterdam, The Netherlands, 2020; pp. 403–426. [CrossRef]

42. Hertel, R. Computational micromagnetism of magnetization processes in nickel nanowires. *J. Magn. Magn. Mater.* **2002**, *249*, 251–256. [CrossRef]

43. Forster, H.; Schrefl, T.; Suess, D.; Scholz, W.; Tsiantos, V.; Dittrich, R.; Fidler, J. Domain wall motion in nanowires using moving grids (invited). *J. Appl. Phys.* **2002**, *91*, 6914. [CrossRef]

44. Ferguson, C.A.; Maclaren, D.A.; McVitie, S. Metastable magnetic domain walls in cylindrical nanowires. *J. Magn. Magn. Mater.* **2015**. [CrossRef]

45. Ivanov, Y.P.; Trabada, D.G.; Chuvilin, A.; Kosel, J.; Chubykalo-Fesenko, O.; Vázquez, M. Crystallographically driven magnetic behaviour of arrays of monocrystalline Co nanowires. *Nanotechnology* **2014**, *25*, 475702. [CrossRef]

46. Talapatra, S.; Tang, X.; Padi, M.; Kim, T.; Vajtai, R.; Sastry, G.V.S.; Shima, M.; Deevi, S.C.; Ajayan, P.M. Synthesis and characterization of cobalt-nickel alloy nanowires. *J. Mater. Sci.* **2009**, *44*, 2271–2275. [CrossRef]

47. Pereira, A.; Palma, J.L.; Vázquez, M.; Denardin, J.C.; Escrig, J. A soft/hard magnetic nanostructure based on multisegmented CoNi nanowires. *Phys. Chem. Chem. Phys.* **2015**. [CrossRef]

48. Liébana Viñas, S.; Salikhov, R.; Bran, C.; Palmero, E.M.; Vazquez, M.; Arvan, B.; Yao, X.; Toson, P.; Fidler, J.; Spasova, M.; et al. Magnetic hardening of $Fe_{30}Co_{70}$ nanowires. *Nanotechnology* **2015**, *26*. [CrossRef]

49. Escrig, J.; Altbir, D.; Jaafar, M.; Navas, D.; Asenjo, A.; Vázquez, M. Remanence of Ni nanowire arrays: Influence of size and labyrinth magnetic structure. *Phys. Rev. B* **2007**, *75*, 184429. [CrossRef]

50. Fernández-Pacheco, A.; Streubel, R.; Fruchart, O.; Hertel, R.; Fischer, P.; Cowburn, R.P. Three-dimensional nanomagnetism. *Nat. Commun.* **2017**, *8*, 15756. [CrossRef] [PubMed]

51. Torrejon, J.; Raposo, V.; Martinez, E.; del Real, R.P.; Hayashi, M. Current-induced dynamics of chiral domain walls in magnetic heterostructures. In *Magnetic Nano- and Microwires*, 2nd ed.; Vázquez, M., Ed.; Elsevier: Amsterdam, The Netherlands, 2020; pp. 297–324. ISBN 978-0-08-102832-2.

52. Méndez, M.; Vega, V.; González, S.; Caballero-Flores, R.; García, J.; Prida, V. Effect of Sharp Diameter Geometrical Modulation on the Magnetization Reversal of Bi-Segmented FeNi Nanowires. *Nanomaterials* **2018**, *8*, 595. [CrossRef] [PubMed]

53. Biziere, N.; Gatel, C.; Lassalle-Balier, R.; Clochard, M.C.; Wegrowe, J.E.; Snoeck, E. Imaging the Fine Structure of a Magnetic Domain Wall in a Ni Nanocylinder. *Nano Lett.* **2013**, *13*, 2053–2057. [CrossRef] [PubMed]

54. Andersen, I.M.; Rodríguez, L.A.; Bran, C.; Marcelot, C.; Joulie, S.; Hungria, T.; Vazquez, M.; Gatel, C.; Snoeck, E. Exotic Transverse-Vortex Magnetic Configurations in CoNi Nanowires. *ACS Nano* **2019**. [CrossRef]

55. Berganza, E.; Bran, C.; Jaafar, M.; Vazquez, M.; Asenjo, A. Domain wall pinning in FeCoCu bamboo-like nanowires. *Sci. Rep.* **2016**, *6*. [CrossRef]

56. Nasirpouri, F.; Peighambari-Sattari, S.-M.; Bran, C.; Palmero, E.M.; Berganza Eguiarte, E.; Vazquez, M.; Patsopoulos, A.; Kechrakos, D. Geometrically designed domain wall trap in tri-segmented nickel magnetic nanowires for spintronics devices. *Sci. Rep.* **2019**, *9*. [CrossRef]

57. Palmero, E.M.; Méndez, M.; González, S.; Bran, C.; Vega, V.; Vázquez, M.; Prida, V.M. Stepwise magnetization reversal of geometrically tuned in diameter Ni and FeCo bi-segmented nanowire arrays. *Nano Res.* **2019**, *12*, 1547–1553. [CrossRef]

58. Pitzschel, K.; Montero Moreno, J.M.; Escrig, J.; Albrecht, O.; Nielsch, K.; Bachmann, J. Controlled Introduction of Diameter Modulations in Arrayed Magnetic Iron Oxide Nanotubes. *ACS Nano* **2009**, *3*, 3463–3468. [CrossRef] [PubMed]

59. Rotaru, A.; Lim, J.-H.; Lenormand, D.; Diaconu, A.; Wiley, J.B.; Postolache, P.; Stancu, A.; Spinu, L. Interactions and reversal-field memory in complex magnetic nanowire arrays. *Phys. Rev. B* **2011**, *84*, 13443. [CrossRef]

60. Arrott, A.S. Visualization and Interpretation of Magnetic Configurations Using Magnetic Charge. *IEEE Magn. Lett.* **2016**, *7*, 1–5. [CrossRef]

61. Ruiz-Gómez, S.; Foerster, M.; Aballe, L.; Proenca, M.P.; Lucas, I.; Prieto, J.L.; Mascaraque, A.; de la Figuera, J.; Quesada, A.; Pérez, L. Observation of a topologically protected state in a magnetic domain wall stabilized by a ferromagnetic chemical barrier. *Sci. Rep.* **2018**, *8*. [CrossRef]

62. Ruiz-Gómez, S.; Fernández-González, C.; Martínez, E.; Raposo, V.; Sorrentino, A.; Foerster, M.; Aballe, L.; Mascaraque, A.; Ferrer, S.; Pérez, L. Helical surface magnetization in nanowires: The role of chirality. *Nanoscale* **2020**, *12*, 17880–17885. [CrossRef]

63. Chizhik, A.; Zhukov, A.; Gonzalez, J.; Gawroński, P.; Kułakowski, K.; Stupakiewicz, A. Spiral magnetic domain structure in cylindrically-shaped microwires. *Sci. Rep.* **2018**, *8*, 15090. [CrossRef]

64. Méndez, M.; González, S.; Vega, V.; Teixeira, J.; Hernando, B.; Luna, C.; Prida, V. Ni-Co Alloy and Multisegmented Ni/Co Nanowire Arrays Modulated in Composition: Structural Characterization and Magnetic Properties. *Crystals* **2017**, *7*, 66. [CrossRef]

. Bran, C.; Berganza, E.; Fernandez-Roldan, J.A.; Palmero, E.M.; Meier, J.; Calle, E.; Jaafar, M.; Foerster, M.; Aballe, L.; Fraile Rodriguez, A.; et al. Magnetization Ratchet in Cylindrical Nanowires. *ACS Nano* **2018**, *12*, 5932–5939. [CrossRef] [PubMed]

. Palmero, E.M.; Bran, C.; Del Real, R.P.; Magén, C.; Vázquez, M. Structural and magnetic characterization of FeCoCu/Cu multilayer nanowire arrays. *IEEE Magn. Lett.* **2014**, *5*, 14–17. [CrossRef]

. Proenca, M.P.; Muñoz, M.; Villaverde, I.; Migliorini, A.; Raposo, V.; Lopez-Diaz, L.; Martinez, E.; Prieto, J.L. Deterministic and time resolved thermo-magnetic switching in a nickel nanowire. *Sci. Rep.* **2019**, *9*. [CrossRef]

. Fernandez-Roldan, J.A.; Del Real, R.P.; Bran, C.; Vazquez, M.; Chubykalo-Fesenko, O. Electric current and field control of vortex structures in cylindrical magnetic nanowires. *Phys. Rev. B* **2020**, *102*, 1–6. [CrossRef]

. Sánchez-Barriga, J.; Lucas, M.; Radu, F.; Martin, E.; Multigner, M.; Marin, P.; Hernando, A.; Rivero, G. Interplay between the magnetic anisotropy contributions of cobalt nanowires. *Phys. Rev. B* **2009**, *80*, 184424. [CrossRef]

. Vivas, L.G.; Escrig, J.; Trabada, D.G.; Badini-Confalonieri, G.A.; Vázquez, M. Magnetic anisotropy in ordered textured Co nanowires. *Appl. Phys. Lett.* **2012**, *100*, 252405. [CrossRef]

. Vivas, L.G.; Yanes, R.; Chubykalo-Fesenko, O.; Vazquez, M. Coercivity of ordered arrays of magnetic Co nanowires with controlled variable lengths. *Appl. Phys. Lett.* **2011**, *98*, 232507. [CrossRef]

. Ivanov, Y.P.; Chuvilin, A.; Lopatin, S.; Kosel, J. Modulated Magnetic Nanowires for Controlling Domain Wall Motion: Toward 3D Magnetic Memories. *ACS Nano* **2016**. [CrossRef] [PubMed]

. Bran, C.; Fernandez-Roldan, J.A.; P Del Real, R.; Asenjo, A.; Chen, Y.-S.; Zhang, J.; Zhang, X.; Fraile Rodríguez, A.; Foerster, M.; Aballe, L.; et al. Unveiling the Origin of Multidomain Structures in Compositionally Modulated Cylindrical Magnetic Nanowires. *ACS Nano* **2020**, *14*. [CrossRef] [PubMed]

. Moreno, J.A.; Mohammed, H.; Kosel, J. Effect of Segment length on domain wall pinning in multisegmented Co/Ni nanowires for 3D memory applications. *J. Magn. Magn. Mater.* **2019**, *484*, 110–113. [CrossRef]

. Berganza, E.; Jaafar, M.; Bran, C.; Fernández-Roldán, J.A.; Chubykalo-Fesenko, O.; Vázquez, M.; Asenjo, A. Multisegmented Nanowires: A Step towards the Control of the Domain Wall Configuration. *Sci. Rep.* **2017**, *7*, 11576. [CrossRef] [PubMed]

 *nanomaterials*

Article

Magnetization Reversal Process and Magnetostatic Interactions in Fe$_{56}$Co$_{44}$/SiO$_2$/Fe$_3$O$_4$ Core/Shell Ferromagnetic Nanowires with Non-Magnetic Interlayer

Javier García [1,*], Alejandro M. Manterola [1], Miguel Méndez [1], Jose Angel Fernández-Roldán [1], Víctor Vega [2], Silvia González [1] and Víctor M. Prida [1,*]

[1] Departamento de Física, Facultad de Ciencias, Universidad de Oviedo, C/Federico García Lorca No. 18, 33007 Oviedo, Spain; alevale71@gmail.com (A.M.M.); miguel.mendez82@gmail.com (M.M.); fernandezroljose@uniovi.es (J.A.F.-R.); gonzalezgana@uniovi.es (S.G.)

[2] Laboratorio de Membranas Nanoporosas, Edificio de Servicios Científico Técnicos "Severo Ochoa", Universidad de Oviedo, C/Fernando Bonguera s/n, 33006 Oviedo, Spain; vegavictor@uniovi.es

* Correspondence: garciafjavier@uniovi.es (J.G.); vmpp@uniovi.es (V.M.P.)

Citation: García, J.; Manterola, A.M.; Méndez, M.; Fernández-Roldán, J.A.; Vega, V.; González, S.; Prida, V.M. Magnetization Reversal Process and Magnetostatic Interactions in Fe$_{56}$Co$_{44}$/SiO$_2$/Fe$_3$O$_4$ Core/Shell Ferromagnetic Nanowires with Non-Magnetic Interlayer. *Nanomaterials* **2021**, *11*, 2282. https://doi.org/10.3390/nano11092282

Academic Editor: George Hadjipanayis

Received: 2 August 2021
Accepted: 31 August 2021
Published: 2 September 2021

Publisher's Note: MDPI stays neutral with regard to jurisdictional claims in published maps and institutional affiliations.

Abstract: Nowadays, numerous works regarding nanowires or nanotubes are being published, studying different combinations of materials or geometries with single or multiple layers. However, works, where both nanotube and nanowires are forming complex structures, are scarcer due to the underlying difficulties that their fabrication and characterization entail. Among the specific applications for these nanostructures that can be used in sensing or high-density magnetic data storage devices, there are the fields of photonics or spintronics. To achieve further improvements in these research fields, a complete understanding of the magnetic properties exhibited by these nanostructures is needed, including their magnetization reversal processes and control of the magnetic domain walls. In order to gain a deeper insight into this topic, complex systems are being fabricated by altering their dimensions or composition. In this work, a successful process flow for the additive fabrication of core/shell nanowires arrays is developed. The core/shell nanostructures fabricated here consist of a magnetic nanowire nucleus (Fe$_{56}$Co$_{44}$), grown by electrodeposition and coated by a non-magnetic SiO$_2$ layer coaxially surrounded by a magnetic Fe$_3$O$_4$ nanotubular coating both fabricated by means of the Atomic Layer Deposition (ALD) technique. Moreover, the magnetization reversal processes of these coaxial nanostructures and the magnetostatic interactions between the two magnetic components are investigated by means of standard magnetometry and First Order Reversal Curve methodology. From this study, a two-step magnetization reversal of the core/shell bimagnetic nanostructure is inferred, which is also corroborated by the hysteresis loops of individual core/shell nanostructures measured by Kerr effect-based magnetometer.

Keywords: nanoporous anodic alumina template; electrodeposition; ALD; magnetic nanowire and nanotube; core/shell nanostructure; FORC analysis; MOKE; magnetization reversal; spintronics

1. Introduction

Cylindrical coaxial nanostructures can exhibit novel phenomena due to their unique size- and shape-dependent physico-chemical properties, as compared to their bulk counterparts, making them enormously attractive as innovative multifunctional materials, for both fundamental and technological applications, such as photonics, drug delivery, and cell separation for proteomics research, hierarchical core/shell heterostructured electrodes for high-performance Li-ion batteries, or highly anisotropic ferromagnetic-antiferromagnetic core/shell nanomaterials exhibiting tunable magnetic properties [1–4]. These peculiar, elongated hybrid nanostructures are coaxially combining two or more components with several phases of distinct properties that could originate additional effects compared with single-phase nanomaterials, which outfit them with enhanced multifunctional properties.

For example, core/shell ferromagnetic nanowires (NWs) and nanotubes (NTs) comprising hybrid metallic/non-metallic core and shell materials usually offer the specific advantage of mixed properties by combining both contributions to convey with tailored multifunctional capabilities, such as the high magnetization and magnetic anisotropy values present in transition metals (TM = Fe, Ni, Co, and their alloys), with the outstanding biocompatibility and enhanced chemical stability exhibited by oxide layers (SiO_2, TiO_2, Fe_3O_4, NiO, CoO, etc.) [5–8].

Several methods have been successfully developed up to date to synthesize these cylindrical coaxial hetero-nanostructures. Among them, the low-cost template-assisted method can be easily combined with other techniques for the fabrication of hybrid core/shell functional nanomaterials, such as electrochemical deposition [9,10], wet-chemical route and coprecipitation [11,12], atomic layer deposition (ALD) [13–17], and chemical vapor deposition (CVD) [18]. For this reason, as well as for its high reproducibility and simplicity, this additive fabrication strategy constitutes one of the most widespread methods for the synthesis of one- and two-dimensional (1D and 2D) coaxial nanostructures. At the same time, three-dimensional (3D) core/shell magnetic nanostructures are nowadays appealing for novel applications in magnetic recording, biotechnology, nanoelectronics, plasmonics and spintronics devices [6,19–22]. Among the vast variety of magnetic heterostructures, bimagnetic core/shell nanocomposites, where both the core and shell couple magnetically hard and soft phases, have recently increased their significance due to the novel physical phenomena they can exhibit. These bimagnetic core/shell nanostructures could efficiently tune their magnetic properties such as the thermal stability of saturation magnetization, high magnetocrystalline anisotropy, and coercivity, through the control of the core/shell parameters, including their shape, size, and chemical composition, showing peculiar and unique features such as large exchange bias, tailored blocking temperatures, tunable coercivities, and stepwise switching of magnetization reversal process [23]. The peculiar features emerging from geometrically curved magnetic nanostructures, including 2D nanowires or 3D curved shells and nanotubes, are strongly related to the size- and shape-dependent effects driven by the interplay between the nanomaterial geometry and topology of each magnetic sub-system [24]. The broad range of emergent physical properties makes these three-dimensional curved nanostructures appealing in view of fundamental research on e.g., skyrmionic systems, photonic and magnonic crystals, or spintronics architectures, also exhibiting these 3D shaped nanomaterials potential applications in photonic, plasmonic and for energy-efficient racetrack magnetic memory devices [25].

Different types of ferromagnetic biphasic nanomaterials have already been reported including soft/hard and hard/soft core/shell bimagnetic nanostructures, respectively [26,27]. However, few works report on three-layered core/shell cylindrical NWs systems recently investigated, where the magnetization switching processes in these hybrid nanostructures remain yet unexplored [13,28–33].

In this study, we investigate the switching process of ferromagnetic/non-magnetic ferrimagnetic core/shell nanocomposite made of $Fe_{56}Co_{44}$(core)/SiO_2(interlayer)/Fe_3O_4(shell). This hybrid core/shell nanocomposite was experimentally synthesized by template-assisted additive fabrication strategy based on nanoporous alumina membrane. The external NT shell of magnetite and the subsequent non-magnetic SiO_2 thin interlayer were deposited by atomic layer deposition (ALD), coupled with the electrochemical deposition technique in order to grow the inner NW core of metallic FeCo alloy. Magneto-optical Kerr effect (MOKE) measurements allow to access the switching process of individual NW and core/shell nanostructures in order to determine the switching of each layer, either the external shell or the inner core of the whole wire/tube nanomaterial. Moreover, the magnetic properties exhibited by these core/shell nanostructures as well as the magnetostatic interactions between the two magnetic contributions of the bimagnetic nanocomposite were investigated by vibrating sample magnetometry (VSM). However, due to the complexity of the system, the core/shell nanowires have to be characterized while they are still embedded in the matrix array. For this purpose, analysis of magnetization reversal curves by means of First Order Reversal Curves (FORC) was also performed for the FeCo NW core, the shell magnetite NT, and the core/shell NW/NT, respectively, with the aim of complementing hysteretic studies regarding the magnetostatic dipolar interaction of the systems, thus allowing us to extract the individual magnetic properties of each nanostructure from the global magnetic behavior of the samples. Our results established that the magnetization of the FeCo core and the magnetite shell may switch at different switching field values during the magnetization reversal of the core/shell NW/NT nanostructure. This confirms the reversal mechanism of our experimental core/shell NWs is uniquely promoted through magnetostatic dipolar coupling between the NT and NW ferromagnetic layers, because of the direct exchange decoupling between both magnetic layers. These results set an appealing strategy for the design of novel 3D magnetic storage and spintronic devices based on core/shell cylindrical NWs that has not been clearly envisaged up to now.

2. Materials and Methods

In this section, all steps for FeCo/Fe$_3$O$_4$ core/shell nanowires fabrication are explained in detail. The complete process flow followed for sample preparation is summarized in Figure 1. First, anodized alumina membranes were fabricated and used as templates for further steps (Figure 1a). The process continues with deposition of the trilayer SiO$_2$/Fe$_2$O$_3$/SiO$_2$ by means of atomic layer deposition, thus forming nanotubes inside the pores of the alumina membranes employed as templates (Figure 1b). After the deposition of a sputtered (and thickened by electrodeposition) Au seed layer on one side of the membrane (Figure 1c), Fe$_{56}$Co$_{44}$ nanowires were electrodeposited inside the former NTs, constituting then the magnetic core of the core/shell nanostructure (Figure 1d). Once the core/shell structure is formed, it is necessary to perform a further thermal treatment to reduce the external layer made of Fe$_2$O$_3$ (hematite) into Fe$_3$O$_4$ (magnetite), therefore, the shell becomes ferrimagnetic (Figure 1e).

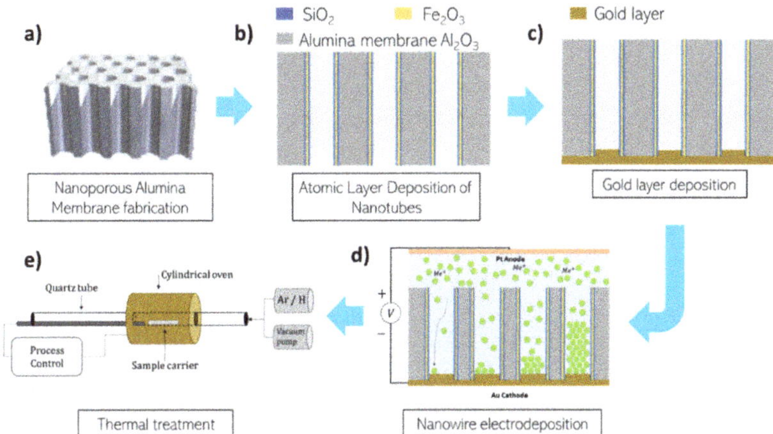

Figure 1. Scheme of the additive fabrication process process to obtain SiO$_2$/Fe$_2$O$_3$/SiO$_2$/FeCo core/shell nanotubes/nanowires, and further thermal treatment for Fe$_2$O$_3$ (hematite) reduction into Fe$_3$O$_4$ (magnetite). (**a**) Nanoporous alumina template, (**b**) deposition of nanotubes shell by ALD, (**c**) gold contact layer deposition by sputtering, (**d**) electrochemical deposition of metallic nanowire core, (**e**) thermal annealing under reducing atmosphere.

2.1. Nanoporous Alumina Membrane Template Fabrication

Hard-Anodized Nanoporous Alumina Membranes (HA-NAMs) were fabricated from electropolished (perchloric acid in ethanol at a rate of 25:75 under an applied voltage of 20 V) high purity aluminum foils (Al 99.999%) using the known process of hard anodiza-

tion [34–36]. The Al substrates were anodized using 0.3 M oxalic acid ($C_2H_2O_4$) at 4 °
for 1.5 h, under an applied voltage of 140 V. After this process, the remaining aluminur
substrate was removed using a solution of 36 g/L of $CuCl_2 \cdot 2H_2O$ and 500 mL/L of HCl
37%. As a final step, the alumina membrane undergoes a process of wet chemical etchir
to open the barrier layer of the pores using 5% wt. phosphoric acid (H_3PO_4) at 30 °C fc
2.5 h leading to a mean pore diameter of 180 nm and a mean interpore distance of 300 nr

2.2. Fabrication of the Nanotubes Shell by Atomic Layer Deposition

The shell of the nanowires, consisting of a three-layered structure is composed c
an external layer of silicon dioxide (SiO_2), a central layer of iron oxide (III) (Fe_2O_3), an
an internal layer of SiO_2, which were grown using the Atomic Layer Deposition (ALI
technique [37].

The silica layers were grown employing three different precursors: (3-aminopropy
triethoxysilane ($H_2N(CH_2)_3 \, Si(OCH_2CH_3)_3$) kept at a temperature of 100 °C; water (H_2C
at 60 °C and ozone (O_3). The reaction temperature in the chamber was fixed to 180 °C. Eac
layer was deposited in a total of 170 ALD cycles, which results in a thickness of aroun
10 nm, according to the estimated deposition rate of 0.06 nm/cycle [38].

The iron oxide (III) layer uses ferrocene ($Fe(C_5H_5)_2$) as precursor along with ozor
with a reaction temperature in the chamber of 230 °C. The layer was deposited after a tot.
of 1150 cycles, resulting in a thickness for the layer of 25 nm, according to the estimate
deposition rate of 0.022 nm/cycle.

2.3. Fabrication of the Nanowires Core by Electrodeposition

The core of the nanowires, composed of an FeCo alloy, is grown by electrochemic.
deposition. Prior to the electrodeposition, a conductive gold layer, which will act as a
anodic electrode during the process, is sputtered and electrodeposited from a commerci.
electrolyte (Orosene 999). The FeCo electrolyte is composed of 0.16 M ($CoSO_4$) + 0.09 M
($FeSO_4$) + 0.06 M ($C_6H_8O_6$) + 0.16 M boric acid (H_3BO_3). The $CoSO_4$, $C_6H_8O_6$, and H_3BC
are mixed together and dissolved in 200 mL of H_2O, heating the dissolution to 40 °
while bubbling it up with N_2. The bubbling helps with the removal of oxygen preser
in the system, after which $FeSO_4$ is added (it would oxidize easily otherwise). After tk
dissolution is completed, it is left to cool down before use. The electrodeposition has bec
performed at a potentiostatic voltage of −1.8V vs. Ag/AgCl reference electrode for 5 mi

In order to recrystallize hematite (Fe_2O_3) in the nanotubes into magnetite (Fe_3C
phase, a new batch of samples underwent a thermal reduction treatment procedure in a fu
nace by annealing at 350 °C in a controlled H_2 (5%) + Ar (95%) atmosphere for 3 h [39,4(
The temperature of the thermal reduction treatment is lower enough to modify the stru
tural and magnetic properties of the FeCo inner core of the core/shell heterostructure.

2.4. Characterization Techniques

Scanning electron microscopy (SEM, JEOL 5600, JEOL, Akishima, Tokyo, Japa:
equipped with an energy dispersive X-ray microanalysis system (EDX, Inca Energy 2C
Oxford Instruments, Abingdon, UK) was employed to perform a morphological and cor
positional characterization of the samples. Transmission electron microscopy (TEM, JEC
JEM 2100, JEOL, Akishima, Tokyo, Japan) suited with an EDX detector (X-MAX, Oxfor
Instruments, Abingdon, UK) and operating in scanning transmission mode was employe
for the EDX linescan analysis of the layered core/shell nanowires hetero-structure.

The magnetic study is performed using a vibrating sample magnetometer (VSM
Versalab, Quantum Design, San Diego, CA, USA) under a magnetic field up to ±3 T ar
temperatures ranging from 50 K to 400 K. Furthermore, hysteresis loops of individu
core/shell nanowires have been measured at room temperature (RT) by means of th
magneto-optical Kerr effect using a NanoMOKE®3 from Durham Magneto Optics L1
(Cambridge, UK).

2.5. First Order Reversal Curve (FORC) Method

FORC methodology is a powerful tool for the study of magnetic properties of materials and it is used to complement the magnetostatic study of the samples that the hysteresis loops would offer by themselves. Although it is not a purely standard characterization technique, the FORC method is increasingly being applied to unravel complex magnetic behaviors. An introduction to this methodology is available in many scientific reports of articles [41,42], reviews [43–46], books [47], and references therein. Very briefly, the idea of this method consists in assuming that the global magnetic hysteresis loop of the sample is composed of different magnetic contributions coming from different sections, elements, or materials that form the sample. The magnetic switching of an individual element is not exclusively dependent on the applied magnetic field but on the magnetic state of its neighboring surroundings which modifies the effective magnetic field that affects it. In order to apply the FORC method, a specific measurement protocol must be followed. After bringing the sample to magnetic saturation, the applied magnetic field is reduced to a certain reversal field H_r, after which the magnetization is recorded along the way back to saturation. Repeating this measurement protocol for several reversal fields (H_r), evenly spaced between positive and negative saturation fields, the FORC distribution $\rho(H_b, H_r)$ can be calculated as the mixed derivative of the magnetization (m) with respect to the reversal field, H_r and the applied magnetic field along the way back to positive saturation, H_b, as shown in Equation (1). This distribution is frequently represented as a contour plot where one can detect a hysteretic process taking place in the H_r, H_b space.

$$\rho(H_b,\ H_r) = -\frac{1}{2}\frac{\partial m}{\partial H_b \partial H_r} \tag{1}$$

3. Results and Discussion

3.1. Samples Fabrication

In order to obtain information regarding the intrinsic magnetic behavior of the different components (magnetic core and magnetic shell) and investigate the different magnetic interactions that may exist among them, different samples were synthesized and studied in this work. All the samples, together with their acronyms are summarized in Table 1. First of all, two HA-NAM templates were coated just with the $SiO_2/Fe_2O_3/SiO_2$ trilayer, corresponding then to those samples where only the nanotube shells are present. The as-deposited (AD) sample does not contain any magnetic phase whilst the sample thermally treated (TT) is formed of ferrimagnetic magnetite nanotubes. In two additional samples, a FeCo core nanowire is electrodeposited inside the NTs of $SiO_2/Fe_2O_3/SiO_2$ coated pores of the alumina membranes. In this case, the as-deposited sample is just the one where only the core is ferromagnetic. After following an appropriate thermal reduction treatment under a controlled atmosphere, both the core and the shell turn into magnetics, being the characterization of this bimagnetic sample the main objective of this study.

Table 1. Designations used for the different samples' terminology employed during this work.

Acronym	Nanotubes (Shell)	Nanowire (Core)	Thermal Treatment	Magnetic	
				Shell	Core
NT-AD	$SiO_2/Fe_2O_3/SiO_2$	None	No	×	×
NT-TT	$SiO_2/Fe_3O_4/SiO_2$	None	Yes (5% H_2 + 95% Ar)	√	×
NWNT-AD	$SiO_2/Fe_2O_3/SiO_2$	FeCo	No	×	√
NWNT-TT	$SiO_2/Fe_3O_4/SiO_2$	FeCo	Yes (5% H_2 + 95% Ar)	√	√

The geometry of the samples has been analyzed by SEM. Figure 2a displays a top view SEM image of the NWNT-AD array, where the different parts that compose the system are identified according to the scheme illustrated in Figure 2b, with an external diameter of the nanotubes shell of around 180 nm and an inner-core of FeCo nanowire with 90 nm in diameter. It is worth mentioning that the apparent elliptical cross-section displayed by the NWNT-AD core/shell nanowires in Figure 2a should be ascribed to artifacts during image acquisition at high magnifications in SEM. The cross-section of the NTNW embedded in the HA-NAM is shown in Figure 2c. The trilayered structure of the core/shell NW can also be observed in the EDX linescan analysis performed by TEM on a single NW and shown in Figure 2d, where the peaks corresponding to the outer iron oxide shell, the SiO_2 interspacing, and the metallic FeCo nucleus are clearly detected. After a deposition time of 5 min, the mean length of the nanowires is around 8 μm. An EDX analysis was performed on a test sample without the deposition of the Fe_2O_3 shell (it would false the nanowires composition otherwise), which confirmed that the average NWs composition was $Fe_{56}Co_{44}$.

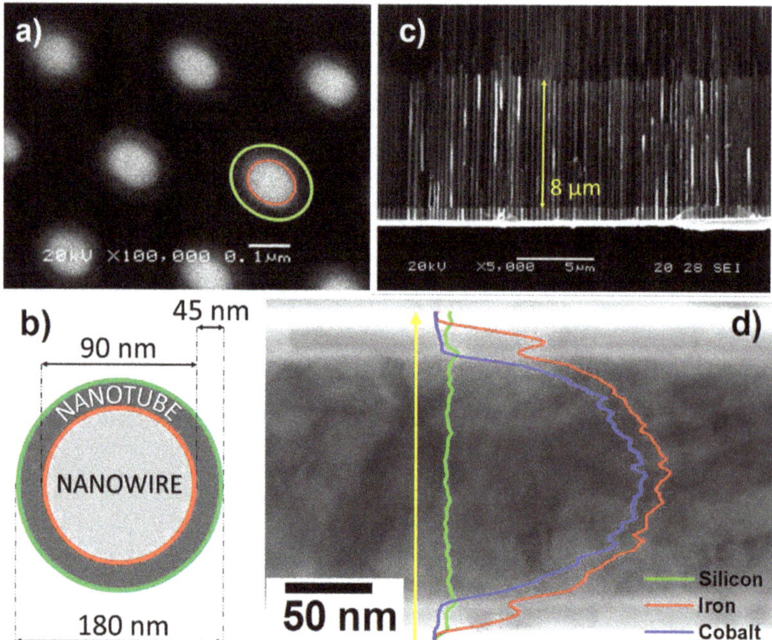

Figure 2. (**a**) Top-view scanning electron micrography of the nanowire/nanotube core/shell system. (**b**) Scheme of the dimensions of the core/shell elements grown in this work. (**c**) Cross-sectional view scanning electron micrography of the nanowire/nanotube system and the mean length of the elements. (**d**) STEM image and EDX linescan analysis performed of a single core/shell NW showing its layered heterostructure.

3.2. Magnetic Properties of the Magnetite Nanotubes

In order to confirm the magnetic phase transformation into magnetite of the hematite nanotubes after the thermal reduction treatment under a controlled atmosphere, a thermomagnetic, M(T), study has been performed. In Figure 3a, the Zero-Field-Cooling, Field Cooling, Field Heating (ZFC-FC-FH) protocol M(T) curves measured at a constant magnetic field of 100 *Oe* is displayed, unmistakably identifying the presence of magnetite phase in the NT shell system due to the detection of the Verwey transition appearing at 125 K [48,49]

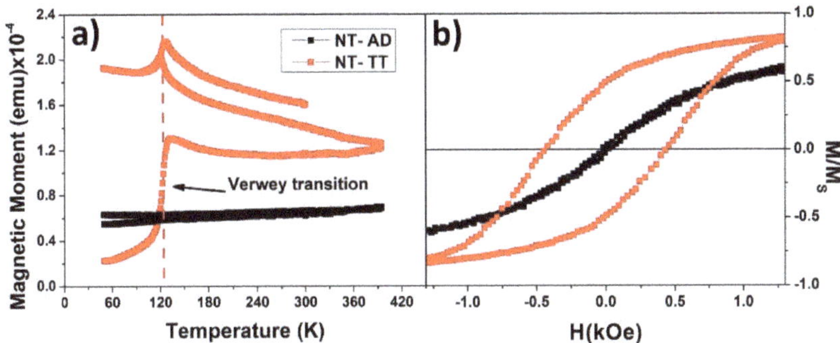

Figure 3. (**a**) ZFC-FC-FH magnetization curves for the shell of Fe_2O_3 NT-AD (black) and Fe_3O_4 NT-TT (red), respectively. (**b**) Normalized VSM hysteresis loops measured by applying the magnetic field along the parallel direction to the hematite (black) and magnetite (red) nanotubes length.

To study the relevance of the shell magnetite phase in the nanotube system, we withdrew some samples from the fabrication process after the ALD procedure but before the nanowires were electrodeposited. It is important to note that for sake of clarity, due to the prevalence of the shape anisotropy for these kinds of systems, the magnetostatic characterization has been restricted to the particular case when applying the external magnetic field along the longitudinal axis of the core/shell nanostructure, which is, in fact, the easy magnetization axis of both, nanowires and nanotubes. A comparison of the magnetic hysteresis loops measured on the nanotube system, before (NT-AD) and after (NT-TT) undergoing the thermal treatment for the hematite phase transformation into magnetite, can be seen in Figure 3b. After performing the thermal treatment under a controlled atmosphere, the resulting shell of magnetite nanotubes has a higher coercivity value, increased from $H_C = 24\ Oe$ for hematite NTs to $H_C = 427\ Oe$ for magnetite NTs.

3.3. Study of the Core/Shell System

The main study of this work has been performed to compare the magnetic behaviour of three samples. We designated the samples in the following way: a first sample which only has the magnetic nanotubes (NT-TT), a second sample with magnetic nanowires but non-magnetic nanotubes (NWNT-AD), and a third sample that has both magnetic nanotubes and nanowires (NWNT-TT). The magnetic behavior of the nanotube and nanowire systems has been characterized individually through the study of the hysteresis loop measured along the axial direction for each sample and then, the corresponding hysteresis loop for the coupled bimagnetic core/shell system was also analyzed.

From the axial hysteresis loops plotted in Figure 4, we can extract a basic idea regarding the magnetic events that take place in each system [50]. By analyzing the slope of the histeretic curves, one can deduce that the NT-TT sample, which only contains the magnetite nanotubes, is a magnetic system where the individual elements which constitute the sample do not magnetically interact so much among them, as its hysteresis loop appears to be straight and relatively parallel to the vertical axis. However, for the NWNT-AD sample, which only contains the $Fe_{56}Co_{44}$ ferromagnetic nanowires, the loop displays a sharper slope than the NT-TT one. This fact indicates that the magnetic elements involved in the system are magnetostatically interacting with each other to a higher degree than in the previous sample. Therefore, the switching fields needed to reverse the magnetization of the magnetic elements of the NWNT-AD sample are altered, making the switching happen at a lower or higher field value (depending on each individual element) than those required to switch them when they are isolated by separately. Lastly, the NWNT-TT sample, which is composed of the core/shell bimagnetic nanostructures, shows an even gentler and steeper slope than the nanowires sample (NWNT-AD), indicating that the two magnetic elements of the core/shell sample turn into even much more interacting between them after the

magnetite nanotubes have been added to the system. The individual magnetic elements
the sample are switching their magnetization state at even lower or higher external fie
values due to the more intense neighboring magnetic field.

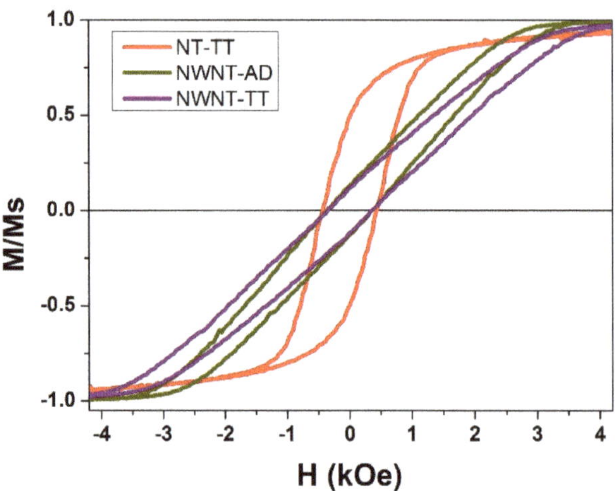

Figure 4. Comparison of the normalized hysteresis loops measured at 300 K for the NT-TT, NWN
AD and NWNT-TT samples by applying the external field along the parallel direction with respect
the nanowire/nanotube longitudinal axis.

In order to further expand this concept, we rely on the more complex FORC metho
ology by obtaining the FORC distributions of the samples at 300 K, all of them measure
by applying the magnetic field parallel to the nanowires/nanotubes long axis. The FOR
distributions for all three samples are collected in Figure 5. Figure 5 (left panel) repr
sents the FORC distribution corresponding to the NT-TT sample which denotes a sing
distribution along the coercivity axis, H_c, thus indicating a weakly interacting syste
as it was also inferred from the corresponding hysteresis loop, where the wide FOR
distribution indicates the broad switching field values of each NT. In contrast to the pr
vious one, the FORC distribution for the NWNT-AD sample, shown in Figure 5 (midd
panel), displays an elongated distribution along the interaction axis, H_u, which indicates
strongly interacting system where each FeCo nanowire influences the magnetic state
its neighbors. Finally, in Figure 5 (right panel) the NWNT-TT FORC distribution displa
a quite similar elongated distribution along the H_u axis to the NWNT-AD, but there a
visible differences in the shape of the distribution. Although apparently, it could resemb
the linear superposition of the two previous FORC distributions, i.e., direct superpositic
of contributions from NWs and NTs, a deeper look into it suggests otherwise. Instead
a FORC diagram comprising the superposition of two independent FORC distribution
it looks that the central distribution along $H_u = 0$, which is in a first instance ascribed
the magnetite nanotubes, shifts the whole magnetization reversal processes of the FeC
nanowires to higher and lower values of the interaction field. This could be explained l
considering that the magnetization reversal of the nanotubes, which occurs completely at
certain value of the applied field, introduces a jump or an offset in the effective magnet
field that affects the whole sample and thus to the nanowires.

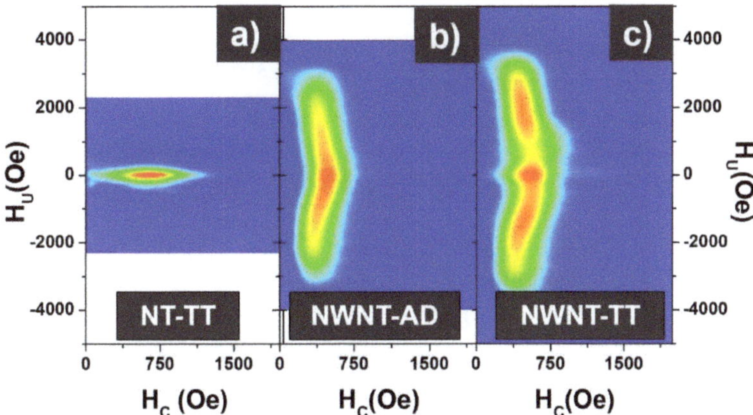

Figure 5. FORC distributions for (**a**) NT-TT sample with only magnetite nanotubes; (**b**) NWNT-AD sample with only FeCo ferromagnetic nanowires; and (**c**) NWNT-TT bimagnetic core/shell sample.

Previous analyses point out that the difference between the nanowire and core/shell systems is the delay of the switching process of the nanowires, represented by a shift towards higher values of the external magnetic field. The presence of the magnetic nanotubes in the core/shell nanowire system, increased the magnetostatic interaction felt by the nanowires, so that the antiparallel coupling between the core and shell is favored. However, magnetostatic interaction among neighboring nanowires is still dominant, as inferred from hysteresis loops and also from FORC analyses. In order to better clarify the magnetization reversal process of individual core/shell nanowires, magneto-optical Kerr effect measurements have been performed with the applied magnetic field parallel to the core/shell longitudinal axis. In Figure 6, the VSM and MOKE parallel hysteresis loops are compared for several samples. The NT-TT VSM and NWNT-AD VSM parallel loops are included as a reference for the magnetization reversal of the nanotubes and the nanowires, respectively. As it can be appreciated from the comparison between the respective VSM and MOKE measurements, the switching field of the single nanowire (NWNT-AD sample with no magnetic shell measured by MOKE, in purple color) matches well with the coercive field value of the FeCo nanowires array measured by VSM (dark yellow color). Furthermore, the blue-colored loop in Figure 6, corresponding with the MOKE hysteresis loop of a single core/shell bimagnetic nanostructure, presents two well-differentiated magnetization jumps. The first one seems to correspond to the ferromagnetic FeCo nanowire in the core, as compared with the one observed for the NWNT-AD sample (purple and dark yellow colors), whilst the second reversal of magnetization is ascribed to the shell of magnetite nanotube, as it is located near the coercive field value of the NT-TT sample obtained from the corresponding VSM hysteresis loop (red color).

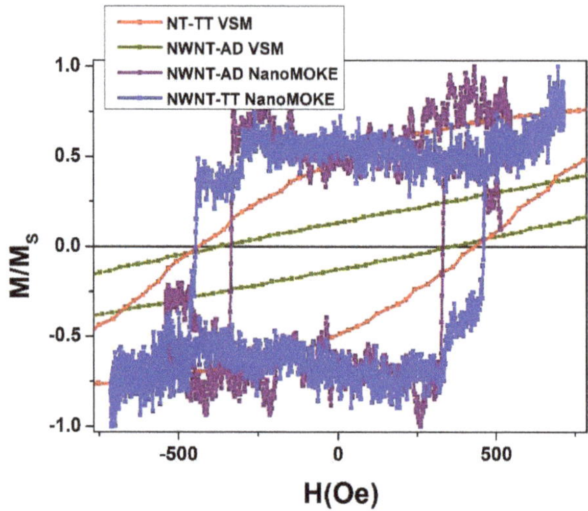

Figure 6. VSM hysteresis loops of the NT-TT (red) and NWNT-AD (dark yellow) arrays. MOK
hysteresis loops of individual $Fe_{56}Co_{44}$ nanowires (purple) and core/shell bimagnetic nanowi
(blue). All hysteresis loops are measured by applying the external field along the parallel directio
with respect to the nanowires/nanotubes longitudinal axis.

4. Conclusions

A cylindrical coaxial bimagnetic core/shell structure composed of $Fe_{56}Co_{44}$ nanowire
and $SiO_2/Fe_2O_3/SiO_2$ multilayered nanotubes were successfully fabricated and magnet
cally characterized. Aiming to design a more complex magnetostatic system, some sample
containing $SiO_2/Fe_2O_3/SiO_2$ nanotubes underwent further thermal treatment in an H
controlled atmosphere (5% H_2 + 95% Ar) to crystallize the hematite (Fe_2O_3) phase int
magnetite (Fe_3O_4) one. From the comparison between the hysteresis loops of the nanotub
nanowire, and nanowire/nanotube systems it was found that the nanotube system doe
not display a highly magnetic interactive behavior while the nanowire system display
higher magnetostatic dipolar interaction between its elements. Additionally, the bima
netic nanowire/nanotube system showed an even higher degree level of magnetostati
interaction among their constituting elements. However, in order to fully understan
the different magnetization processes occurring in the samples, the First-Order Revers
Curves methodology was introduced. The FORC analysis performed on the shell nanotub
samples showed a weakly magnetostatic interaction system, whereas the nanowire an
nanowire/nanotube samples' magnetic behavior was dictated by a system with strongl
magnetostatic interacting elements. A deeper analysis of the core/shell nanowire/nanotub
sample shows evidence of the magnetostatic coupling of dipolar origin between the tw
magnetic parts, the core, and the shell, of the bimagnetic nanostructure, in such a wa
that their magnetic interactions cause the delay in the switching of their magnetic state a
compared to their individual magnetic behaviors separately. Finally, the two-step switcl
ing of the magnetization reversal process for the bimagnetic nanowire/nanotube syster
was confirmed by MOKE measurements performed on several individual elements of th
core/shell sample.

Author Contributions: Conceptualization, V.M.P. and J.G.; methodology, V.V., S.G. and M.M.; va
idation, V.M.P., J.G. and V.V.; formal analysis, J.G. and J.A.F.-R.; investigation, A.M.M., M.M. an
S.G.; resources, V.M.P. and J.G.; data curation, J.G. and J.A.F.-R.; writing—original draft preparatio
A.M.M., J.G. and V.V.; writing—review and editing, J.G. and V.M.P.; visualization, V.V., S.G. and J.C
supervision, J.G. and V.M.P.; project administration, V.V. and J.G.; funding acquisition, V.M.P. and J.C
All authors have read and agreed to the published version of the manuscript.

Funding: This research was funded by Spanish Ministerio de Ciencia e Innovación (MICINN) and Research Agency State (AEI), under grant number PID2019-108075RB-C32.

Acknowledgments: Authors would like to acknowledge the technical support supplied by the Scientific-Technical services of the University of Oviedo. Javier García acknowledges the financial support from the Principality of Asturias and European Union under the PCTI/FEDER program "Retorno Talento" (project No. IDI/2018/000010). Spanish MICINN is also gratefully recognized under research program FEDER-11-UNOV10-2E-1011.

Conflicts of Interest: The authors declare no conflict of interest.

References

Lauhon, L.J.; Gudiksen, M.S.; Wang, D.; Lieber, C.M. Epitaxial core-shell and core-multishell nanowire heterostructures. *Nature* **2002**, *420*, 57–61. [CrossRef] [PubMed]

Chen, H.; Deng, C.; Zhang, X. Synthesis of Fe_3O_4@SiO_2@PMMA Core-Shell-Shell Magnetic Microspheres for Highly Efficient Enrichment of Peptides and Proteins for MALDI-ToF MS Analysis. *Angew. Chem. Int. Ed.* **2010**, *49*, 607–611. [CrossRef] [PubMed]

Zhang, Q.; Chen, H.; Wang, J.; Xu, D.; Li, X.; Yang, Y.; Zhang, K. Growth of Hierarchical 3D Mesoporous $NiSi_x$/NiCo2O4 Core/Shell Heterostructures on Nickel Foam for Lithium-Ion Batteries. *ChemSusChem* **2014**, *7*, 2325–2334. [CrossRef]

Irfan, M.; Wang, C.J.; Khan, U.; Li, W.J.; Zhang, X.M.; Kong, W.J.; Liu, P.; Wan, C.H.; Liu, Y.W.; Han, X.F. Controllable synthesis of ferromagnetic–antiferromagnetic core-shell NWs with tunable magnetic properties. *Nanoscale* **2017**, *9*, 5694–5700. [CrossRef] [PubMed]

Romero, V.; Vega, V.; García, J.; Zierold, R.; Nielsch, K.; Prida, V.M.; Hernando, B.; Benavente, J. Changes in Morphology and Ionic Transport Induced by ALD SiO_2 Coating of Nanoporous Alumina Membranes. *ACS Appl. Mater. Interfaces* **2013**, *5*, 3556–3564. [CrossRef]

Gong, M.; Kirkeminde, A.; Skomski, R.; Cui, J.; Ren, S. Template-Directed FeCo Nanoshells on AuCu. *Small* **2014**, *10*, 4118–4122. [CrossRef]

Ivanov, Y.P.; Alfadhel, A.; Alnassar, M.; Perez, J.E.; Vazquez, M.; Chuvilin, A.; Kosel, J. Tunable magnetic nanowires for biomedical and harsh environment applications. *Sci. Rep.* **2016**, *6*, 24189. [CrossRef]

Zamani Kouhpanji, M.R.; Um, J.; Stadler, B.J.H. Demultiplexing of Magnetic Nanowires with Overlapping Signatures for Tagged Biological Species. *ACS Appl. Nano Mater.* **2020**, *3*, 3080–3087. [CrossRef]

Chen, J.Y.; Ahmad, N.; Shi, D.W.; Zhou, W.P.; Han, X.F. Synthesis and magnetic characterization of Co-NiO-Ni core-shell nanotube arrays. *J. Appl. Phys.* **2011**, *110*, 73912. [CrossRef]

Shi, D.; Chen, J.; Riaz, S.; Zhou, W.; Han, X. Controlled nanostructuring of multiphase core–shell nanowires by a template-assisted electrodeposition approach. *Nanotechnology* **2012**, *23*, 305601. [CrossRef]

Zhou, X.; Feng, W.; Qiu, K.; Chen, L.; Wang, W.; Nie, W.; Mo, X.; He, C. BMP-2 Derived Peptide and Dexamethasone Incorporated Mesoporous Silica Nanoparticles for Enhanced Osteogenic Differentiation of Bone Mesenchymal Stem Cells. *ACS Appl. Mater. Interfaces* **2015**, *7*, 15777–15789. [CrossRef] [PubMed]

Li, X.; Li, H.; Song, G.; Peng, Z.; Ma, L.; Meng, C.; Liu, Y.; Ding, K. Preparation and Magnetic Properties of Nd/FM (FM=Fe, Co, Ni)/PA66 Three-Layer Coaxial Nanocables. *Nanoscale Res. Lett.* **2018**, *13*, 326. [CrossRef] [PubMed]

Chong, Y.T.; Görlitz, D.; Martens, S.; Yau, M.Y.E.; Allende, S.; Bachmann, J.; Nielsch, K. Multilayered Core/Shell Nanowires Displaying Two Distinct Magnetic Switching Events. *Adv. Mater.* **2010**, *22*, 2435–2439. [CrossRef] [PubMed]

Pitzschel, K.; Bachmann, J.; Montero-Moreno, J.M.; Escrig, J.; Görlitz, D.; Nielsch, K. Reversal modes and magnetostatic interactions in Fe_3O_4/ZrO_2/Fe_3O_4 multilayer nanotubes. *Nanotechnology* **2012**, *23*, 495718. [CrossRef] [PubMed]

Méndez, M.; Vega, V.; González, S.; Caballero-Flores, R.; García, J.; Prida, V. Effect of Sharp Diameter Geometrical Modulation on the Magnetization Reversal of Bi-Segmented FeNi Nanowires. *Nanomaterials* **2018**, *8*, 595. [CrossRef]

Fernandez-Roldan, J.; Chrischon, D.; Dorneles, L.; Chubykalo-Fesenko, O.; Vazquez, M.; Bran, C. A Comparative Study of Magnetic Properties of Large Diameter Co Nanowires and Nanotubes. *Nanomaterials* **2018**, *8*, 692. [CrossRef]

Palmero, E.M.; Méndez, M.; González, S.; Bran, C.; Vega, V.; Vázquez, M.; Prida, V.M. Stepwise magnetization reversal of geometrically tuned in diameter Ni and FeCo bi-segmented nanowire arrays. *Nano Res.* **2019**, *12*, 1547–1553. [CrossRef]

Mu, C.; Yu, Y.-X.; Wang, R.M.; Wu, K.; Xu, D.S.; Guo, G.-L. Uniform Metal Nanotube Arrays by Multistep Template Replication and Electrodeposition. *Adv. Mater.* **2004**, *16*, 1550–1553. [CrossRef]

Li, C.-L.; Chang, C.-J.; Chen, J.-K. Fabrication of sandwich structured devices encapsulating core/shell SiO_2/Fe_3O_4 nanoparticle microspheres as media for magneto-responsive transmittance. *Sens. Actuators B Chem.* **2015**, *210*, 46–55. [CrossRef]

Wang, F.; Wang, N.; Han, X.; Liu, D.; Wang, Y.; Cui, L.; Xu, P.; Du, Y. Core-shell FeCo@carbon nanoparticles encapsulated in polydopamine-derived carbon nanocages for efficient microwave absorption. *Carbon N. Y.* **2019**, *145*, 701–711. [CrossRef]

Moon, S.H.; Noh, S.; Lee, J.-H.; Shin, T.-H.; Lim, Y.; Cheon, J. Ultrathin Interface Regime of Core–Shell Magnetic Nanoparticles for Effective Magnetism Tailoring. *Nano Lett.* **2017**, *17*, 800–804. [CrossRef] [PubMed]

Zhang, W.; Yang, W.; Chandrasena, R.U.; Özdöl, V.B.; Ciston, J.; Kornecki, M.; Raju, S.; Brennan, R.; Gray, A.X.; Ren, S. The effect of core–shell engineering on the energy product of magnetic nanometals. *Chem. Commun.* **2018**, *54*, 11005–11008. [CrossRef] [PubMed]

23. Al Salmi, W.; Gyawali, P.; Dahal, B.; Pegg, I.L.; Philip, J. Core-shell FeNi-NixFe3-xO4 nanowires. *J. Vac. Sci. Technol. B* **2015**, 40604. [CrossRef]
24. Streubel, R.; Fischer, P.; Kronast, F.; Kravchuk, V.P.; Sheka, D.D.; Gaididei, Y.; Schmidt, O.G.; Makarov, D. Magnetism in curved geometries. *J. Phys. D Appl. Phys.* **2016**, *49*, 363001. [CrossRef]
25. Fernández-Pacheco, A.; Streubel, R.; Fruchart, O.; Hertel, R.; Fischer, P.; Cowburn, R.P. Three-dimensional nanomagnetism. *Nat. Commun.* **2017**, *8*, 15756. [CrossRef]
26. Shamaila, S.; Liu, D.P.; Sharif, R.; Chen, J.Y.; Liu, H.R.; Han, X.F. Electrochemical fabrication and magnetization properties of CoCrPt nanowires and nanotubes. *Appl. Phys. Lett.* **2009**, *94*, 203101. [CrossRef]
27. Kim, H.W.; Na, H.G.; Kwak, D.S.; Kwon, Y.J.; Van Khai, T.; Lee, C.; Jung, J.H. Fabrication and magnetic properties of In$_2$O$_3$/NiMnGa core-shell nanowires. *Thin Solid Films* **2013**, *546*, 219–225. [CrossRef]
28. Otálora, J.A.; López-López, J.A.; Vargas, P.; Landeros, P. Chirality switching and propagation control of a vortex domain wall in ferromagnetic nanotubes. *Appl. Phys. Lett.* **2012**, *100*, 72407. [CrossRef]
29. Zhang, J.; Agramunt-Puig, S.; Del-Valle, N.; Navau, C.; Baró, M.D.; Estradé, S.; Peiró, F.; Pané, S.; Nelson, B.J.; Sanchez, A.; et al. Tailoring Staircase-like Hysteresis Loops in Electrodeposited Trisegmented Magnetic Nanowires: A Strategy toward Minimization of Interwire Interactions. *ACS Appl. Mater. Interfaces* **2016**, *8*, 4109–4117. [CrossRef] [PubMed]
30. Hertel, R. Ultrafast domain wall dynamics in magnetic nanotubes and nanowires. *J. Phys. Condens. Matter* **2016**, *28*, 48300 [CrossRef]
31. Chen, A.P.; Gonzalez, J.; Guslienko, K. Magnetization Reversal Modes in Short Nanotubes with Chiral Vortex Domain Wall. *Materials* **2018**, *11*, 101. [CrossRef]
32. Proenca, M.P.; Rial, J.; Araujo, J.P.; Sousa, C.T. Magnetic reversal modes in cylindrical nanostructures: From disks to wires. *Sci. Rep.* **2021**, *11*, 10100. [CrossRef]
33. Magén, C.; Pablo-Navarro, J.; De Teresa, J.M. Focused-Electron-Beam Engineering of 3D Magnetic Nanowires. *Nanomaterials* **2021**, *11*, 402. [CrossRef]
34. Lee, W.; Ji, R.; Gösele, U.; Nielsch, K. Fast fabrication of long-range ordered porous alumina membranes by hard anodization. *Nat. Mater.* **2006**, *5*, 741–747. [CrossRef]
35. Vega, V.; Böhnert, T.; Martens, S.; Waleczek, M.; Montero-Moreno, J.M.; Görlitz, D.; Prida, V.M.; Nielsch, K. Tuning the magnetic anisotropy of Co-Ni nanowires: Comparison between single nanowires and nanowire arrays in hard-anodic aluminum oxide membranes. *Nanotechnology* **2012**, *23*, 465709. [CrossRef]
36. Vega, V.; García, J.; Montero-Moreno, J.M.; Hernando, B.; Bachmann, J.; Prida, V.M.; Nielsch, K. Unveiling the Hard Anodization Regime of Aluminum: Insight into Nanopores Self-Organization and Growth Mechanism. *ACS Appl. Mater. Interfaces* **2015**, 28682–28692. [CrossRef] [PubMed]
37. George, S.M. Atomic Layer Deposition: An Overview. *Chem. Rev.* **2010**, *110*, 111–131. [CrossRef] [PubMed]
38. Bachmann, J.; Zierold, R.; Chong, Y.T.; Hauert, R.; Sturm, C.; Schmidt-Grund, R.; Rheinländer, B.; Grundmann, M.; Gösele, U.; Nielsch, K. A Practical, Self-Catalytic, Atomic Layer Deposition of Silicon Dioxide. *Angew. Chem. Int. Ed.* **2008**, *47*, 6177–6179. [CrossRef] [PubMed]
39. Matthews, A. Magnetite forrnation by the reduction of hematite with iron under hydrothermal conditions. *Am. Mineral.* **1976**, 927–932.
40. Zierold, R.; Le Lam, C.; Dendooven, J.; Gooth, J.; Böhnert, T.; Sergelius, P.; Munnik, F.; Montero Moreno, J.M.; Görlitz, D.; Detavernier, C.; et al. Magnetic characterization and electrical field-induced switching of magnetite thin films synthesized by atomic layer deposition and subsequent thermal reduction. *J. Phys. D Appl. Phys.* **2014**, *47*, 485001. [CrossRef]
41. Valdez-Grijalva, M.A.; Muxworthy, A.R. First-order reversal curve (FORC) diagrams of nanomagnets with cubic magnetocrystalline anisotropy: A numerical approach. *J. Magn. Magn. Mater.* **2019**, *471*, 359–364. [CrossRef]
42. Ruta, S.; Hovorka, O.; Huang, P.W.; Wang, K.; Ju, G.; Chantrell, R. First order reversal curves and intrinsic parameter determination for magnetic materials; Limitations of hysteron-based approaches in correlated systems. *Sci. Rep.* **2017**, *7*, 45218. [CrossRef] [PubMed]
43. Roberts, A.P.; Heslop, D.; Zhao, X.; Pike, C.R. Understanding fine magnetic particle systems through use of first-order reversal curve diagrams. *Rev. Geophys.* **2014**, *52*, 557–602. [CrossRef]
44. Pike, R. First-order reversal-curve diagrams and reversible magnetization. *Phys. Rev. B-Condens. Matter Mater. Phys.* **2003**, *68*, 1- [CrossRef]
45. Pike, C.R.; Ross, C.A.; Scalettar, R.T.; Zimanyi, G. First-order reversal curve diagram analysis of a perpendicular nickel nanopillar array. *Phys. Rev. B* **2005**, *71*, 134407. [CrossRef]
46. Groß, F.; Ilse, S.E.; Schütz, G.; Gräfe, J.; Goering, E. Interpreting first-order reversal curves beyond the Preisach model: An experimental permalloy microarray investigation. *Phys. Rev. B* **2019**, *99*, 064401. [CrossRef]
47. Beron, F.; Carignan, L.-P.; Menard, D.; Yelon, A. Extracting Individual Properties from Global Behaviour: First-order Reversal Curve Method Applied to Magnetic Nanowire Arrays. In *Electrodeposited Nanowires and Their Applications*; Lupu, N., Ed.; InTech: Rijeka, Croatia, 2010.
48. Bohra, M.; Agarwal, N.; Singh, V. A Short Review on Verwey Transition in Nanostructured Fe$_3$O$_4$ Materials. *J. Nanomater.* **2019**, *2019*, 8457383. [CrossRef]

9. Walz, F. The Verwey transition-a topical review. *J. Phys. Condens. Matter* **2002**, *14*, R285–R340. [CrossRef]
0. Martínez Huerta, J.M.; De La Torre Medina, J.; Piraux, L.; Encinas, A. Self consistent measurement and removal of the dipolar interaction field in magnetic particle assemblies and the determination of their intrinsic switching field distribution. *J. Appl. Phys.* **2012**, *111*, 083914. [CrossRef]

45

 *nanomaterials*

Article

Spin Caloritronics in 3D Interconnected Nanowire Networks

Tristan da Câmara Santa Clara Gomes [†] , **Nicolas Marchal** [†] **and Flavio Abreu Araujo** [†] **and Luc Piraux** *[*,†]

Institute of Condensed Matter and Nanosciences, Université Catholique de Louvain, Place Croix du Sud 1, 1348 Louvain-la-Neuve, Belgium; tristan.dacamara@uclouvain.be (T.d.C.S.C.G.); nicolas.marchal@uclouvain.be (N.M.); flavio.abreuaraujo@uclouvain.be (F.A.A.)

* Correspondence: luc.piraux@uclouvain.be

† These authors contributed equally to this work.

Received: 24 September 2020; Accepted: 19 October 2020; Published: 22 October 2020

Abstract: Recently, interconnected nanowire networks have been found suitable as flexible macroscopic spin caloritronic devices. The 3D nanowire networks are fabricated by direct electrodeposition in track-etched polymer templates with crossed nano-channels. This technique allows the fabrication of crossed nanowires consisting of both homogeneous ferromagnetic metals and multilayer stack with successive layers of ferromagnetic and non-magnetic metals, with controlled morphology and material composition. The networks exhibit extremely high, magnetically modulated thermoelectric power factors. Moreover, large spin-dependent Seebeck coefficients were directly extracted from experimental measurements on multilayer nanowire networks. This work provides a simple and cost-effective way to fabricate large-scale flexible and shapeable thermoelectric devices exploiting the spin degree of freedom.

Keywords: 3D nanowire networks; spin caloritronics; thermoelectricity; spintronics; giant magnetoresistance multilayers

---

1. Introduction

Spin-dependent transport mechanisms are expected to play a crucial role in the development of next generation of thermoelectric devices [1]. Therefore, a central focus of the rapidly emerging field of spin caloritronics is combining heat-driven transport with spintronics [2,3]. Previous studies on nanoscale metal structures, magnetic tunnel junctions and magnetic insulators have led to the observation of various spin-enabled mechanisms that may differ significantly from conventional thermoelectrics effects, including spin Seebeck effects [4,5], thermally driven spin injection [6] and thermal assisted spin-transfer torque [7,8]. However, dimensions in magnetic nanostructures lead to major experimental issues such as insufficient power output capability and lack of reliable methods to obtain key spin caloritronic parameters, and have limited the application of spin caloritronic devices based on these effects [9,10].

In this context, recently developed interconnected magnetic nanowire (NW) networks embedded within porous polymer films provide a simple and cost-effective pathway to fabricate flexible and shapeable, macroscopic-scale spintronic nanoarchitectures with advantageous thermoelectric properties. They combine the macroscopic dimension required for large thermoelectric power output with the nanostructuration required to enable spintronic effects, and meet the mechanical, electrical and thermal stability required for practical applications. Indeed, ferromagnetic transition metals exhibit large diffusion thermopowers because of the pronounced structure of the d-band and the high energy derivative of the density of states at the Fermi level [11], while they also exhibit significant magnon-drag contribution to the thermoelectric power within a wide temperature range [11,12], leading to both positive and negative large Seebeck coefficient S at room temperature (RT). Moreover, due to their large electrical conductivity σ, they can exhibit very large power factors PF $= S^2\sigma$, which is the physical parameter that relates to the output power density of a thermoelectric material, that remain in interconnected NW structures made of ferromagnetic metals. Therefore, interconnected magnetic NW networks allow to obtain both p-type and n-type light, robust, flexible and shapeable thermoelectric elements, which are both required for practical thermometric devices. Besides, interconnected networks made of multilayered NWs, with a succession of ferromagnetic metal (FM) and normal metal (NM) layers, allow for large magnetic control of thermoelectric transport and for the precise and direct extraction of spin-dependent Seebeck coefficients from experimental measurements. Therefore, such NW-based spin caloritronics devices overcome the insufficient power generation capability inherent to the custom-patterned nanoscale magnetic structures reported previously and constitute promising candidates for heat management applications [13,14].

2. Materials and Methods

Three-dimensional (3D) nanoporous templates were obtained by a track-etched technique with a sequential multi-step exposure of energetic heavy ions, at various angles with respect to the normal of polycarbonate (PC) film surface [15,16]. It allows to obtain 20 µm thick template films with distinct porosities and pores sizes, as illustrated in Figure 1a. In the present study, the as-prepared polymer membranes containing networks of interconnected cylindrical nanopores were designed with pores of well-defined diameter of about 80 nm and with porosity $P \approx 3\%$. Then, the PC membranes were filled with 3D metallic NWs by direct electrodeposition at RT in the potentiostatic mode using a Ag/AgCl reference electrode and a Pt counter electrode, which allows for a very high degree of replication of the nanopores [16–18]. For this, the PC templates were coated on one side with a Au/Cr bilayer using an e-beam evaporator to serve as cathode during the electrochemical deposition. Interconnected homogeneous NW networks, as illustrated in Figure 1b, made of several ferromagnetic pure metals (Ni, Co) and alloys (NiFe, CoNi, NiCr, CoCr) were grown by electrodeposition from home-made electrolyte solutions with fine-tuned metallic ions concentration and pH acidity at calibrated constant potentials [16–20]. In addition, FM and Cu layers were electrochemically stacked in the host 3D porous templates to make interconnected FM/Cu multilayered NW networks, with FM = Co, $Co_{50}Ni_{50}$, $Ni_{80}Fe_{20}$ and Ni, as illustrated in Figure 1c. They were grown by electrodeposition from single home-made electrolyte solution using a pulsed electrodeposition technique [13,14,21,22].

The interconnected NW structure was characterized using a field-emission scanning electron microscope (FE-SEM) after complete chemical dissolution of the cathode and the template. The crossed NW networks are mechanically robust and self-standing, as shown by Figure 1d,e where SEM images of the networks are presented. As seen, the NW networks exhibit the replicated complex branching morphology of the porous template. The inset in Figure 1d shows the typical size as well as the mechanical robustness of the macroscopic self-supporting networks. In order to perform magnetoresistance (MR) and magneto-thermoelectric power (MTP) measurements on the interconnected NWs embedded into the PC film (schematically shown in Figure 2a), the Au cathode was locally etched by plasma to pattern two Au electrodes with low contact resistance, as illustrated in Figures 2b–d. The Seebeck coefficient was measured by inducing a temperature difference ΔT using a resistive element and measuring the induced thermoelectric voltage ΔV, following the procedure described in refs. [13,22]. MR and MTP were measured by applying an external magnetic field along the out-of-plane (OOP) and in-plane (IP) directions of the NW network films, with a maximum applied field of ±10 kOe. The measurements were made at temperatures from 10 K to 320 K. Other benefits of the 3D interconnected nanowire network system are its light weight and flexibility. For instance, the density of Co NWs fully filling a PC membrane was estimated to be about 1.4 g/cm^3. The flexibility of the network films is shown in Figure 2e. As seen, the film can be easily twisted without damaging its electrical properties. The electrical resistance was measured during several successive torsions of a network of interconnected nanowires, revealing only a marginal variation in resistance, less than 0.2%.

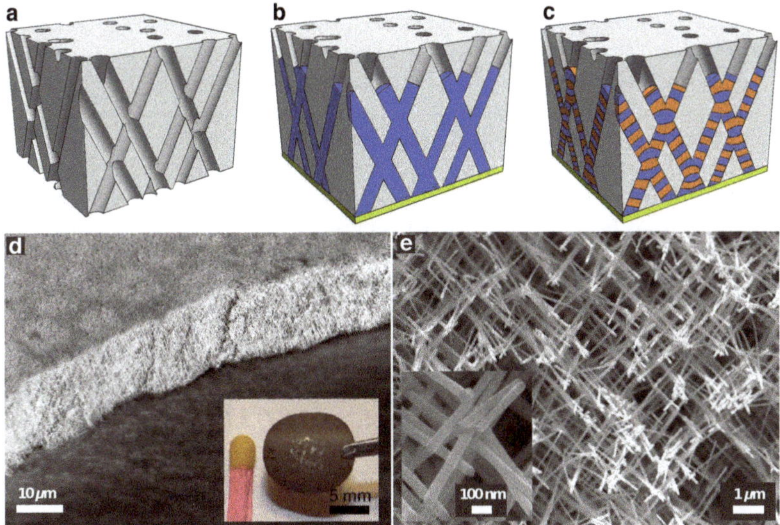

Figure 1. (**a**) Schematics of the 3D nanoporous polymer template, (**b**) crossed nanowire and (**c**) crossed nanowire network with alternating magnetic and non-magnetic layers. (**d**,**e**) SEM images of self-supported interconnected nanowire network with different magnifications. (**d**) Low-magnification image showing the 50° tilted view of a macroscopic nanowire network film with 105 nm diameter and ∼20% packing density. The inset displays an optical image showing the size and mechanical robustness of the macroscopic self-supporting network. (**e**) Low magnification image showing the top view of the NW network with 80 nm diameter and ∼3% packing density. The inset shows the nanowire branched structure at higher magnification.

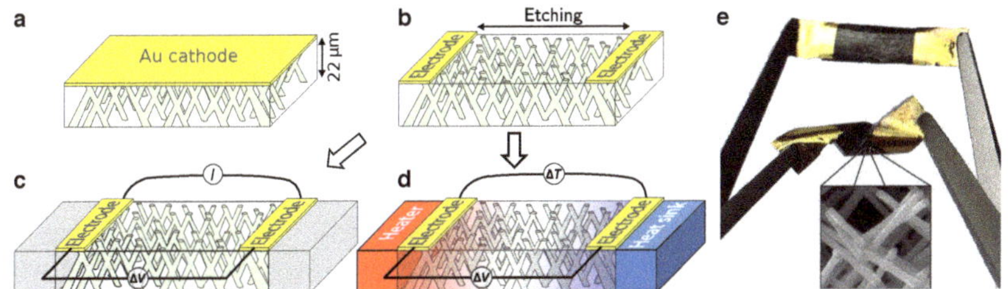

Figure 2. 3D interconnected nanowire networks and experimental set-ups for measurement of transport properties. (**a**) Schematic of 3D interconnected nanowire network film grown by electrodeposition from a Au cathode into a 20 µm thick polycarbonate template with crossed-nanopores. (**b**) Two-probe electrodes design obtained by local etching of the Au cathode. (**c**,**d**) Device configuration for successive measurements of the resistance and the Seebeck coefficient. (**c**) The voltage differential ΔV induced by the injected current I between the two metallic electrodes is measured while the two electrodes are maintained at an identical and constant temperature. (**d**) Heat flow is generated by a resistive element at one electrode while the other electrode is maintained at desired temperature. The temperature difference ΔT between the two metallic electrodes is measured by a thermocouple while thermoelectric voltage ΔV settles. (**e**) Photograph of a flexible device made of 3D interconnected nanowires embedded in a polycarbonate matrix and with the two gold electrodes design. The inset SEM image shows the nanowire branched structure with diameter of 80 nm and ∼3% packing density.

3. Results and Discussion

3.1. Homogeneous Nanowire Networks

The resistance and Seebeck coefficient of homogeneous NW networks were measured for $T = 10$–320 K. Table 1 provides the Seebeck coefficients of homogeneous NW networks made of several metals and alloys at RT. Both n-type and p-type thermoelectric NW networks with large absolute thermopower values were obtained. The results are consistent with reported bulk Seebeck coefficients [23–26]. Moreover, the value reported for the $Co_{50}Ni_{50}$ NW network is also consistent with the ones reported for parallel arrays of NiCo NWs [27,28]. The resistivity of the homogeneous NW networks were estimated assuming that the Matthiesen's rule holds for the different metallic NW networks. In that case, the resistivity at RT is given by $\rho_{NWs}^{RT} = \rho_{FM}^{RT} + \rho_{NWs}^{0}$, where ρ_{FM}^{RT} is the resistivity of the FM that composes the NWs at RT due to thermally excited scatterings and ρ_{NWs}^{0} is the residual resistivity of the NWs due to impurities along with surface scattering within the NW network and internal grain-boundary scattering. For NW diameter not too small ($\phi \geq 40$ nm), the thermally induced scattering effects are independent on the sample dimensions, nanostructuration and defect concentration [29]. Therefore, ρ_{FM}^{RT} can be taken as the ideal resistivity values at RT reported for bulk materials in the literature (from refs. [30–33]). Moreover, because the resistivity due to thermally excited scatterings tends to 0 at low temperatures, the resistivity at 10 K of the NW networks can be approximated to $\rho_{NWs}^{10\,K} \approx \rho_{NWs}^{0}$. Finally, using the residual resistivity ratio $RRR = R_{NWs}^{RT}/R_{NWs}^{10\,K} \approx (\rho_{FM}^{RT} + \rho_{NWs}^{0})/\rho_{NWs}^{0}$, the RT resistivity of the NWs can be estimated as $\rho_{NWs}^{RT} \approx \rho_{FM}^{RT} RRR/(RRR - 1)$. The calculated resistivity are provided in Table 1. Slightly larger resistivity compared to the bulk materials have been obtained, as expected for electrodeposited nanostructured materials. Indeed, electrodeposited materials display a relatively large amount of defects, leading to defect scattering into the NWs, while the NW transverse nanoscale dimensions lead to surface scattering effect. This engenders larger residual resistivity of the NW networks compared to bulk materials.

Table 1 also provides the power factors $PF = S^2/\rho$ together with the figure of merit ZT for the NW networks at RT. Due to the slightly larger electrical resistivity of the NW networks compared to bulk materials, slightly lower PF values have been found with respect to the bulk values. However, the PF values obtained are similar an even larger than to that of widely used thermoelectric material bismuth-telluride (in the range 1–6 mW/K^{2}m) and at least one order of magnitude larger than the ones reported for flexible thermoelectric films based on optimized conducting polymers [34,35]. Large RT value of PF of about 11.0 mW/K^{2}m have been obtained in interconnected Co NWs, which is almost as good as the bulk Co that display the largest PF value of about 15 mW/K^{2}m [36]. Moreover, regarding p-type materials, the RT PF estimated for the Fe and $Ni_{96}Cr_4$ NW networks are close to the largest PF at RT of about 9 mW/K^{2}m found for $CePd_3$ [37,38]. Therefore, electrodeposited magnetic NW networks are suitable for both n-type and p-type thermoelectric modules, in particular for active cooling applications as it has been recently shown that such application requires materials exhibiting large PF [38,39]. The efficiency of a material's thermoelectric energy conversion is determined by its figure of merit $ZT = S^2\sigma T/\kappa$, with κ the thermal conductivity. In a previous study, M. Ou et al [40] have measured the thermal conductivity of a suspended Ni NW for $T = 15$–300 K. While the Lorenz ratio $L = \kappa/\sigma T$ departs from the Sommerfeld value ($L_0 = 2.45 \cdot 10^{-8}$ V^2/K^2) at low temperatures, L was found to be equal to L_0 with a 5% margin of error above $T = 50$ K. Due to the very low thermal conductivity of polycarbonate ($\kappa = 0.2$ W/Km at RT), the contribution of the polymer matrix to heat transport is much smaller than that of the metallic NW network. Indeed, assuming that the Wiedemann–Franz law holds for the NWs, estimations of the RT electronic thermal conductivities $\kappa_e = L_0 T/\rho$ provides values between 10 and 100 W/Km, hence at least two orders of magnitude above the thermal conductivity of the PC template. Moreover, in the limits of the Wiedemann–Franz law, the ZT value can be approximated by $ZT \approx S^2/L_0$. Using this approximation, the ZT values of the NWs networks at RT have been estimated and are reported in Table 1. Although the figure of merit is more than one order of magnitude smaller than those of state-of-the-art thermoelectric materials ($ZT \approx 1$ in BiTe alloys), it is comparable to those of thermocouple alloys ($ZT \approx 6 \cdot 10^{-2}$ and $ZT \approx 1.4 \cdot 10^{-2}$ in constantan and chromel, respectively) and can be used in applications for devices with low energy requirements when the supply of heat essentially is free as with waste heat.

Table 1. Room-temperature Seebeck coefficient S, resistivity ρ, power factor (PF), figure of merit ZT, magnetoresistance ratio (MR) and magnetothermopower ratio (MTP) of interconnected homogeneous nanowire networks made of ferromagnetic metals and alloys.

	S (μV/K)	ρ ($\mu\Omega$cm)	PF (mW/K^{2}m)	ZT (-)	MR (%)	MTP (%)
Co	−28.0	7.1	11.0	$3.2 \cdot 10^{-2}$	1.1	−1.1
Fe	+15.0	12.8	1.8	$9.1 \cdot 10^{-3}$	0.2	-
Ni	−19.6	9.1	4.2	$1.6 \cdot 10^{-2}$	1.6	−6.0
Co$_{50}$Ni$_{50}$	−18.3	15.4	2.2	$1.4 \cdot 10^{-2}$	2.9	−3.8
Ni$_{90}$Fe$_{10}$	−34.4	18.6	6.3	$4.8 \cdot 10^{-2}$	3.3	−2.1
Ni$_{80}$Fe$_{20}$	−36.9	25.0	5.4	$5.6 \cdot 10^{-2}$	2.4	−1.5
Ni$_{70}$Fe$_{30}$	−40.2	32.5	5.0	$6.6 \cdot 10^{-2}$	1.5	−1.3
Ni$_{60}$Fe$_{40}$	−46.0	42.4	5.0	$8.6 \cdot 10^{-2}$	0.7	−0.8
Ni$_{96}$Cr$_4$	+15.5	27.3	0.9	$9.8 \cdot 10^{-3}$	0.1	-

Besides, magneto-transport measurements have been conducted. Figure 3a–h shows the variation of resistance and Seebeck coefficient with a magnetic field applied along the IP and OOP direction for the Co (a–b), Co$_{50}$Ni$_{50}$ (c–d), Ni$_{80}$Fe$_{20}$ (e–f) an Ni (g–h) NW networks. The MR curves show the anisotropic magnetoresistance (AMR) effect that leads to a decrease in resistivity as the angle between the magnetization and current directions is increased. Indeed, the current flow being restricted along the NW segments, the saturation magnetization in the IP direction makes an average angle of ±65° with the current. By comparison, when the magnetization is saturated in the OOP direction, the average angle between the magnetization and the current directions is much smaller (±25°), leading to a larger resistance at saturation when the field is applied in the OOP direction than when the field is applied in the IP direction. The lower resistance state expected for the perpendicular configuration between magnetization and current could not be achieved in such NW networks due to their geometry. Figure 3a–h also shows that the absolute value of the thermopower increases with increasing angle between the magnetization and the current flow for the NW networks. At saturation, larger absolute thermopower is obtained in the IP direction compared to the OOP direction. This is in good agreement with previous studies performed on single NWs [27]. Moreover, as shown in Figure 3a–h for all NW networks, the MR and MTP effects exhibit similar behavior, indicating a direct relation between the two effects, as expected from Mott's formula for the diffusion thermopower. Table 1 provides the RT values of the MR and MTP ratio, defined as MR $= (R_{\max} - R_{\min})/R_{\max}$ and MTP $= (S_{\max} - S_{\min})/S_{\max}$, respectively. Similar amplitudes have been obtained for the Co, Co$_{50}$Ni$_{50}$ and NiFe alloys networks. The slightly larger MTP ratio amplitude compared to the MR ratio amplitude observed in Co$_{50}$Ni$_{50}$ is also consistent with previous studies performed on NiCo alloy NWs [27,28]. In contrast, Figure 3g,h reveals an enhancement of the MTP ratio up to three times larger than the corresponding MR ratio amplitude for the Ni NW network, in spite of similar field dependencies. Such definition of the MR ratios leads to an underestimation of the traditional AMR ratio defined as $(R_{\parallel} - R_{\perp})/R_{\perp}$, where $R_{\parallel}$ and $R_{\perp}$ are the theoretical resistance states obtained for parallel and perpendicular orientations of the magnetization and current directions, respectively. However, the AMR ratio can be extracted from the resistance states measured at saturation in the OOP and IP directions R_{OOP} and R_{IP} using the analytical model described in refs. [17–19].

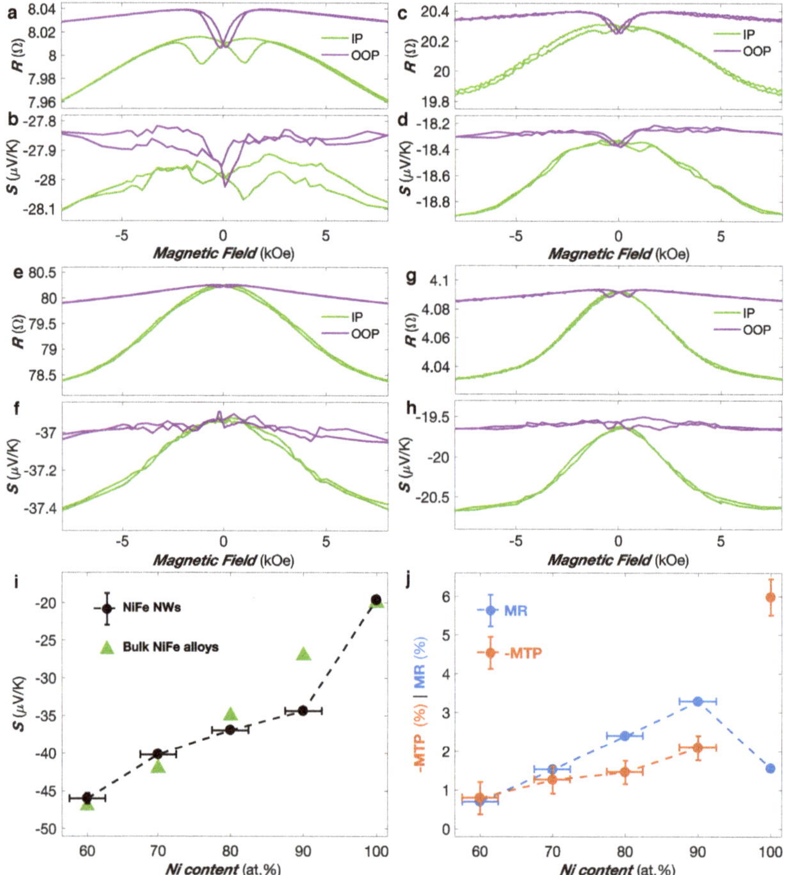

Figure 3. (**a–h**) Room-temperature variation of the electrical resistance and Seebeck coefficient of Co (**a,b**), $Co_{50}Ni_{50}$ (**c,d**), $Ni_{80}Fe_{20}$ (**e,f**) and Ni (**g,h**) nanowire networks obtained with the magnetic field applied along the in-plane (IP—green) and out-of-plane (OOP—purple) directions of the nanowire network film. (**i**) Variation of the Seebeck coefficient vs Ni content in NiFe nanowire networks at room temperature. Values previously reported for bulk alloys [41] are also shown. (**j**) Magnetoresistance (MR—blue) and magneto-thermopower (MTP—red) ratios as a function of Ni content in NiFe nanowire networks at room temperature.

Figure 3i shows the Seebeck coefficients S at RT of the NW networks made of Ni_xFe_{1-x} alloys with $0.6 \leq x \leq 1$. As seen, the thermopower increases continuously with increasing Fe content, reaching values between $-20\ \mu V/K$ for pure Ni to about $-45\ \mu V/K$ for $Ni_{60}Fe_{40}$. These results are in good agreement with the experimental data obtained on bulk NiFe alloys [24,41]. Therefore, NiFe alloys with fine-tune composition potentially yield significantly larger Seebeck coefficients than pure ferromagnetic metals like Co and thermocouple materials like constantan ($Cu_{55}Ni_{45}$: $S \approx -38\ \mu V/K$). The measured value for $Ni_{80}Fe_{20}$ NWs ($S \approx -37\ \mu V/K$) is also very similar to the reported bulk values in the literature [25,26]. Figure 3j shows the magnitude of the MR and MTP ratios evaluated at RT for pure Ni and Ni_xFe_{1-x} alloy NW networks as a function of the Ni content x. It reveals a peak in the MR ratio for alloying compositions around 90% of Ni, in coherence with previous studies on bulk NiFe alloys [42,43]. Moreover, it highlights the very different behavior of the Ni NW network, compared to the NiFe alloys. For the Ni_xFe_{1-x} alloy samples with $0.6 \leq x \leq 0.9$, the magnitude of the MTP ratio is either comparable or smaller to the MR ratio. The smaller value of the MTP ratio with respect to the corresponding MR ratio for the $Ni_{80}Fe_{20}$ NW network is in agreement with measurements performed on $Ni_{80}Fe_{20}$ thin films [44]. In contrast, the Ni NW network exhibits a MTP effect of -6% much larger than the MR ratio of $\sim1.6\%$. This result is in good agreement with previous measurements performed on single Ni NWs and parallel arrays of Ni NWs, where MTP ratios were found up to 2.5–3 times larger than the corresponding MR ratios [27,28,45] and may be related to the spin-dependent Seebeck coefficients, $S_\uparrow$ and $S_\downarrow$, of opposite sign [46]. It is interesting to note that for Ni thin films, the observed anisotropic MTP has approximately the same magnitude than the anisotropic MR ($\sim1.5\%$) [44]. Further studies are needed to understand this unexpected enhanced MTP for Ni NWs.

The addition of transition metal impurities in ferromagnetic metals has been found to have a large influence on their spin-dependent electronic transport properties (see ref. [47] for a review). It can be ascribed to a matching/mismatching of the d-electronic states between the host ferromagnetic metal and the transition metal impurities [48]. Notably, diluted Cr impurities in Co, Fe and Ni generate stronger scattering in the majority-spin channel in a two-band model. This leads to a spin-asymmetry parameter $\alpha = \rho_\downarrow/\rho_\uparrow < 1$, which means that minority spin electrons dominate the electronic conduction, contrasting to that of pure ferromagnetic metals. In this context, NW networks made of dilute NiCr and CoCr alloys (Cr content ≤ 7 at.%) have been fabricated, and the influence of impurity concentration on the AMR effect and thermopower has been investigated [20]. Figure 4a provides the MR ratio as a function of the Cr content for NiCr NW networks at RT and $T = 100$ K, where the MR ratio is defined as MR $= (R_{OOP} - R_{IP})/R_{OOP}$, with R_{OOP} and R_{IP} the resistance states reached at $H = 10$ kOe in the OOP and IP directions, respectively. Although this MR ratio underestimates the AMR effect, its sign correctly reflects the sign of the AMR effect. Indeed, when the magnetization is saturated along the OOP and IP directions, it makes respectively average angles of about $25°$ and $65°$ with the current flow that is strictly restricted along the NW segments. Therefore, the case $R_{OOP} > R_{IP}$ unambiguously indicates a positive AMR effect, as usually observed for ferromagnetic metals and alloys. As seen in Figure 4a, positive AMR have been found at RT for all interconnected NiCr NW networks, with a decrease of the MR ratio with increasing Cr content. Conversely, negative AMR effect at $T = 100$ K has been observed for NiCr NW networks with Cr content $> 4\%$. This is illustrated in the inset of Figure 4a that compares the MR curves measured at RT and $T = 11$ K for the $Ni_{96}Cr_4$ sample. While the AMR is positive at RT, the $Ni_{96}Cr_4$ NW network exhibits $R_{IP} > R_{OOP}$ at $T = 11$ K. Similar results have been observed on the NiCr NWs samples with Cr content $> 4\%$. These observations are in agreement with previous negative AMR measurements reported at low temperatures in bulk NiCr dilute alloys [49–51].

Figure 4b shows the MR curves for the dilute CoCr NW networks, where negative AMR have been observed at RT for Cr concentrations ≥ 3 at.% contrasting to pure Co NWs [20]. Indeed, as seen in the inset of Figure 4b, the RT MR curves measured for the Co and $Co_{95}Cr_5$ samples show opposite behaviors ($R_{OOP} > R_{IP}$ and $R_{IP} > R_{OOP}$ for Co and $Co_{95}Cr_5$ NW networks, respectively). Moreover, all CoCr NWs exhibit negative AMR with similar amplitude for all alloying composition down to ~1 at.%. Even if the interpretation is still controversial [52], Campbell, Fert and Jaoul have proposed a model based on a spin-orbit mechanism [50,53] predicting that the AMR ratio is proportional to $(\alpha - 1)$ at low temperatures, where $\alpha = \rho_\downarrow^0/\rho_\uparrow^0$ is the spin asymmetry coefficient with $\rho_\downarrow^0$ and $\rho_\uparrow^0$ the residual resistivities of the spin down and spin up electrons. When $\alpha < 1$, negative AMR is thus expected. Moreover, in this simplified model, α is constant for a given impurity, independently of its concentration. When the temperature increases, the electron-phonon and electron-magnon scattering processes induces additional terms in the electrical resistivity. In consequence, as temperature rises, the AMR ratio of the dilute alloy should tend to that of the host pure ferromagnetic metal, converging more rapidly as the impurity concentration is lower [53]. The predictions of Campbell, Fert and Jaoul [50,53] are mostly in qualitatively good agreement with the experimental results obtained for both interconnected NiCr and CoCr NW networks and the theoretic prediction of spin asymmetry coefficient $\alpha < 1$ in these dilute alloys [47,54]. Notably, the negative MR ratios observed at low temperatures for CoCr NW networks are found to be independent of the Cr content, while the change of sign at RT is favored for the lowest Cr concentration (see Figure 4b). However, the fact that the MR ratio remains negative for Cr content ≥ 3 at.% in CoCr NWs is unexpected because thermally activated scattering of conduction electrons should significantly weaken the effect. Similar unexpected negative AMR ratios at RT have already been observed in earlier studies performed on bulk CoIr alloys [52]. One hypothetical justification for the RT negative MR ratio in the CoCr NW networks is that the residual resistivity due to impurity scatterings which is proportional to the Cr content becomes dominant compared to the thermally activated resistivity of the matrix at RT in CoCr alloys with Cr content $\geq 3\%$. Thus their influence over the AMR effect remain dominant at RT. This is illustrated by the residual ratio ($R_{300\,K}/R_{10\,K}$) below 2 (1.7 and 1.6 for 3 and 5 at.%Cr respectively) for the samples with RT negative AMR, up to 2 for the sample with positive RT AMR (5.4 and 2.7 for the Co and $Co_{99}Cr_1$ samples, respectively).

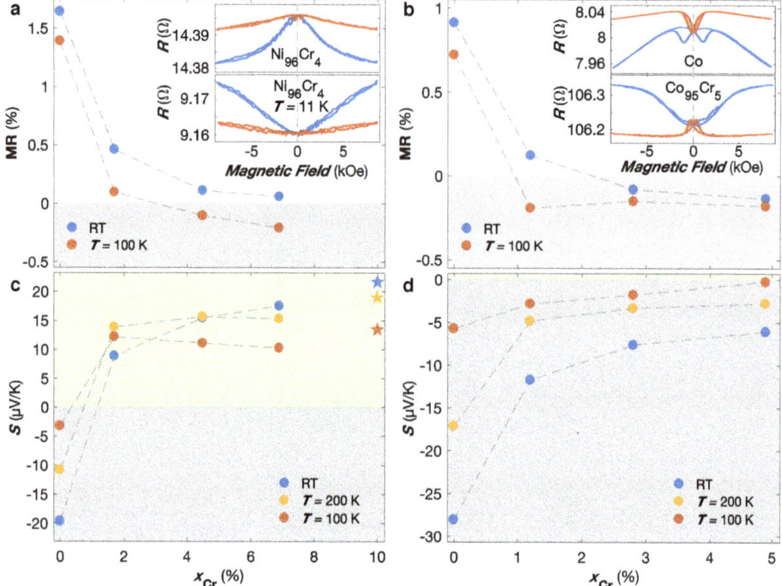

Figure 4. (**a,b**) Magnetoresistance (MR) ratio as a function of the Cr content for interconnected NiCr (**a**) and CoCr (**b**) nanowire networks at room temperature (RT—blue) and $T = 100$ K (red). The gray areas in (**a,b**) indicates negative anisotropic magnetoresistance. The inset in (**a**) compares the MR curves measured along the out-of-plane (red) and in-plane (blue) directions for the $Ni_{96}Cr_4$ crossed nanowire network at RT and $T = 11$ K. The inset in (**b**) compares the RT MR curves measured along the out-of-plane (red) and in-plane (blue) directions for the interconnected Co and $Co_{95}Cr_5$ nanowire networks. (**c,d**) Seebeck coefficient S at zero magnetic field as a function of the Cr content of for interconnected NiCr (**c**) and CoCr (**d**) nanowire networks at RT (blue), $T = 200$ K (yellow) and $T = 100$ K (red). The recommended values for chromel ($Ni_{90}Cr_{10}$) are indicated by star symbols. The green and gray areas in (**c,d**) indicate positive and negative S values, respectively. The symbol size encompasses the experimental error bars, which are set to two times the standard deviation of the experimental measurements of the electrical and temperature measurements, gathering 95% of the data variation.

Figure 4c,d show the Seebeck coefficient as a function of the Cr concentration in the absence of any external magnetic field for the interconnected NiCr and CoCr NW networks at RT, $T = 200$ K and $T = 100$ K. As seen in Figure 4c, the thermopower of NiCr NWs abruptly changes sign, going from the pure Ni value (-20 µV/K) to about $+17.5$ µV/K for the $Ni_{93}Cr_7$ NW network [20]. Similar changes of sign in the Seebeck coefficient have been previously observed in bulk NiCr alloys with the electrical conduction dominated by minority spin electrons and have been related to drastic modifications in the density of states at the Fermi level by the addition of Cr impurities to the host ferromagnetic metals [46,47,55]. Figure 4c also provides the recommended values for the thermocouple material Chromel P ($Ni_{90}Cr_{10}$), which are consistent with the Seebeck coefficients reported for the studied samples. In contrast, the sign of the Seebeck coefficient does not change for any value of Cr concentration for CoCr NW networks, as seen in Figure 4d. The RT thermopower drops rapidly from -28 µV/K for pure Co to much smaller negative values approaching -5 µV/K for $Co_{95}Cr_5$ NWs. Similar features are also observed at lower temperatures. These observations are consistent with experiments previously conducted on dilute Co-based alloys with $\alpha < 1$ [11]. The sudden drop of S induced by the incorporation of few at. % of Cr in Co NWs can also be ascribed to significant changes in the density of states for the majority spin electrons. Please note that extremely small MTP effects (typically less than 1%) were found in all NiCr and CoCr samples.

3.2. Multilayered Nanowire Networks

We will now turn to the discussion to interconnected FM/Cu multilayered NW networks, which have been found to be good candidates for spin caloritronic devices with large spin-dependent thermoelectric transport properties as they combine large giant magnetoresistance (GMR) effect measured in the current-perpendicular-to-plane (CPP) configuration with large thermoelectric power factors [13,14,22]. In addition, their macroscopic dimension leads to efficient energy conversion and opens the door to practical applications. In this context, interconnected FM/Cu multilayered NW networks, with FM = Co, $Co_{50}Ni_{50}$ or $Ni_{80}Fe_{20}$ and Ni, have been investigated. In CPP FM/Cu multilayers, the Seebeck coefficient along the axial direction (perpendicular direction to the layers) can be calculated from the corresponding transport properties using Kirchhoff's rules [56] as

$$S_{\perp} = \frac{S_{Cu}\kappa_{FM} + \lambda S_{FM}\kappa_{Cu}}{\lambda\kappa_{Cu} + \kappa_{FM}}. \tag{1}$$

Here S_{FM}, S_{Cu} and κ_{FM}, κ_{Cu} represent the thermopower and the thermal conductivity of the FM and Cu and $\lambda = t_{FM}/t_{Cu}$ the thickness ratio of FM and Cu layers. If the Wiedemann–Franz law holds for the Cu and FM, Equation (1) simply reduces to

$$S_{\perp} = \frac{S_{Cu}\rho_{Cu} + \lambda S_{FM}\rho_{FM}}{\lambda\rho_{FM} + \rho_{Cu}}, \tag{2}$$

with ρ_{FM} and ρ_{Cu} being the corresponding electrical resistivities. According to Equation (2), in the limits $S_{FM}\rho_{FM} \gg S_{Cu}\rho_{Cu}$, which is usually acceptable, and a thickness ratio λ not too small, $S_{FM/Cu}$ is mainly determined by the large thermopower of the ferromagnetic metal. The same conclusion can be drawn when the FM layers are made of ferromagnetic alloys such as $Co_{50}Ni_{50}$ and $Ni_{80}Fe_{20}$, because of highly contrasting thermal conductivity values between these alloys and Cu. By comparison, the Seebeck coefficient of a planar FM/Cu multilayer stack in the direction parallel to the layers (CIP) is given by

$$S_{\parallel} = \frac{S_{Cu}\rho_{FM} + \lambda S_{FM}\rho_{Cu}}{\lambda\rho_{Cu} + \rho_{FM}}. \tag{3}$$

It shows that large thermopowers can be obtained only if the thickness ratio λ is very large, although in CIP system with dominant interface scattering, the thermopower becomes less sensitive to the Cu bulk resistance, which leads to larger S value [57]. This contrasting behavior between the Seebeck coefficient in the CIP and CPP configurations is illustrated in Figure 5 for Co/Cu (a) and $Ni_{80}Fe_{20}$/Cu (b) multilayers, using Equations (2) and (3) and literature room-temperature resistivity and thermopower values for bulk Co, $Ni_{80}Fe_{20}$ and Cu [23,25,30,41,58]. Although the electrical resistivity and thermal conductivity values for multilayered nanowires may vary significantly compared with their respective bulk constituents, the same trends in the contrasting behavior between the two configurations remain. So, CPP FM/Cu multilayers are promising candidates for good thermoelectric materials.

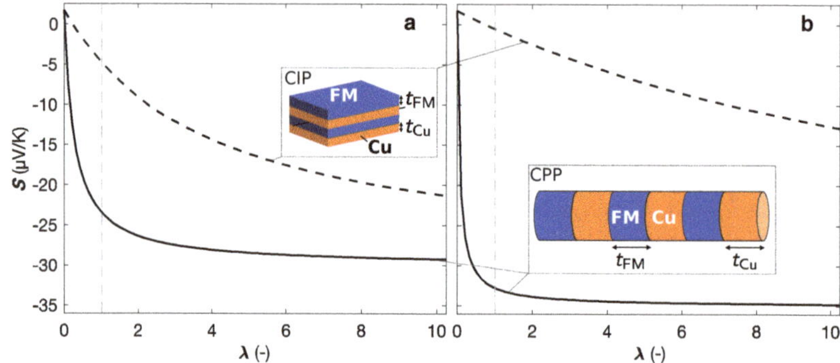

Figure 5. Calculated thermopowers for (**a**) Co/Cu and (**b**) Ni$_{80}$Fe$_{20}$/Cu multilayers in the layer parallel (dash-doted line—CIP) and perpendicular (solid line—CPP) directions as a function of the thickness ratio $\lambda = t_{FM}/t_{Cu}$, using Equations (2) and (3) and the bulk values for S_{FM}, ρ_{FM}, S_{Cu} and ρ_{Cu}, with (**a**) FM = Co and (**b**) FM = Ni$_{80}$Fe$_{20}$. The grey dashed line shows the values for $\lambda = 1$. The insets in (**a**,**b**) show FM/Cu multilayer stacks with CIP (**a**) and CPP (**b**) configurations.

Figure 6 shows the resistance and thermopower variation with a magnetic field for the FM/Cu NW samples, with FM = Co (a–b), Co$_{50}$Ni$_{50}$ (c–d), Ni$_{80}$Fe$_{20}$ (e–f) and Ni (g–h). As shown in Figure 6a,b, the resistance and thermopower of the Co/Cu NW sample show the same magnetic field dependencies and similar relative changes of ~25% at $H = \pm 8$ kOe at RT. The value of S in the saturated state of the Co/Cu NWs of about -20 μV/K is only slightly lower than the one reported for homogeneous Co NW networks (~-28 μV/K), in good agreement with Equation (2) and Figure 5a. In addition, the sample is nearly magnetically isotropic, as observed from the magneto-transport curves obtained with the applied magnetic field along the OOP and IP directions. This behaviour corresponds to the one expected considering the crossed NW architecture and magneto-static arguments when using similar FM and NM layer thicknesses [59]. Similar features have been observed on the Co$_{50}$Ni$_{50}$/Cu and Ni$_{80}$Fe$_{20}$/Cu NW networks. As seen in Figure 6c,d, the resistance and thermopower of the Co$_{50}$Ni$_{50}$/Cu sample show the same magnetic field dependencies and relative changes of ~30% at $H = \pm 8$ kOe at RT. Slightly higher Seebeck coefficients have been recorded in the Co$_{50}$Ni$_{50}$/Cu NW network, with a Seebeck coefficient in the saturated state in the IP direction of about -22 μV/K. These values are in agreement with the results obtained from measurements carried out on single Co$_{50}$Ni$_{50}$ and Co$_{50}$Ni$_{50}$/Cu NWs [27,60]. Figure 6e,f shows the RT resistance and thermopower of the Ni$_{80}$Fe$_{20}$/Cu NW network. Despite very similar magnetic field dependencies, the relative changes in Seebeck coefficient (~26%) is found to be larger than the one of the resistance (~17%) at $H = \pm 8$ kOe. As expected, the measured RT thermopower on the CPP-GMR Ni$_{80}$Fe$_{20}$/Cu sample in the saturated state (~-25 μV/K) is only slightly smaller than the value found in the homogeneous Ni$_{80}$Fe$_{20}$ NW network (~-35 μV/K), in good agreement with Equation (1) and Figure 5b. In contrast, the RT Seebeck coefficients reported for Ni$_{80}$Fe$_{20}$/Cu multilayers in the CIP geometry (~-10 μV/K) are much smaller [61]. A contrasting behavior has been observed in interconnected Ni/Cu NW networks. Figure 6g,h shows the RT resistance and thermopower field dependencies with the applied magnetic field along the IP and OOP directions of the Ni/Cu NW network. Despite similar field dependencies, the amplitude of the magneto-thermopower effect (MTP $\approx -3.7\%$) is about four times larger than the corresponding magnetoresistance effect (MR $\approx 0.8\%$) at RT. This larger value of the MTP with respect to the MR ratio is consistent with our result obtained in homogeneous Ni NWs and previous studies on Ni/Cu multilayers [62]. This significant boost of the MTP, compared to MR may indicate underlying contrasting $S_\uparrow$ and $S_\downarrow$ values. This is consistent with the values of $S_\uparrow = -43$ μV/K and $S_\downarrow = 20$ μV/K reported by Cadeville and Roussell [46].

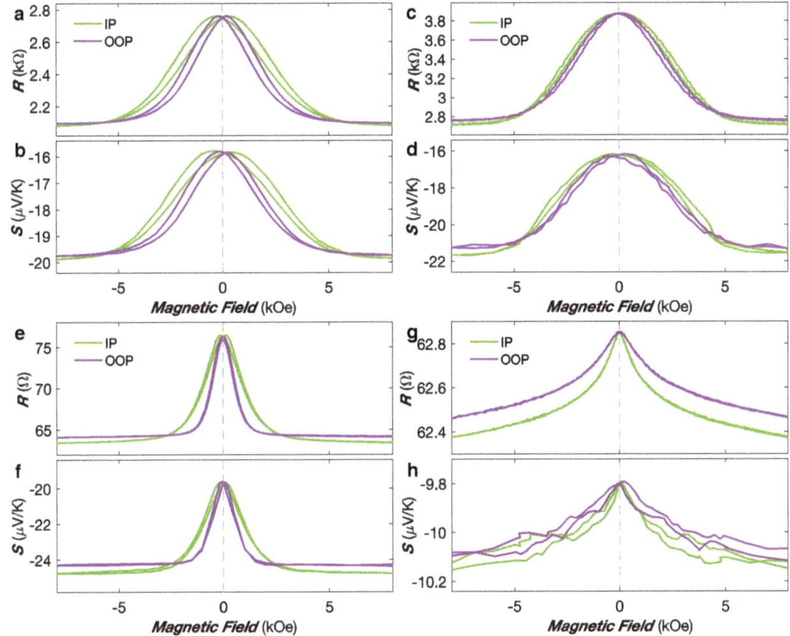

Figure 6. (a–h) Room-temperature variation of the electrical resistance and Seebeck coefficient of Co/Cu (**a,b**), Co$_{50}$Ni$_{50}$/Cu (**c,d**), Ni$_{80}$Fe$_{20}$/Cu (**e,f**) and Ni/Cu (**g,h**) multilayered nanowire samples obtained with the applied field in-plane (IP—green) and out-of-plane (OOP—purple) of the NW network film.

Table 2 provides the Seebeck coefficient in the saturated states (S_P) for the different FM/Cu NW network together with the estimated electrical resistivity (ρ_P) and the thermoelectric power factor (PF$_P$) at RT (see [13,14] for more details about the resistivity calculation). It shows that the CPP geometry of the device is suitable for spin caloritronic purposes since the RT thermopower and power factor of the FM/Cu NW networks in the saturated state are only slightly smaller than the value found in the pure corresponding ferromagnetic materials, as expected from conventional rules for a stacking arrangement in series (see Equation (2)). The figure of merit of the FM/Cu NW networks was estimated by $ZT = S^2/L_0$ at RT [13,14] and is reported in the saturated state in Table 2. The values found are comparable to those of homogeneous NW networks. Furthermore, considering the energy conversion from heat to electric power, the ZT of the proposed device is much larger than that of a spin Seebeck power generator ($ZT \approx 10^{-4}$) based on a device using a two-step conversion process and the inverse spin Hall effect to convert spin current to charge current in non-magnetic materials [63]. Finally, Table 2 also provides the values at RT of the magnetoresistance ratio MR $= (R_{AP} - R_P)/R_{AP}$, with R_{AP} and R_P being respectively the high- and low-resistance states, and the absolute value of the magneto-thermopower MTP $= (S_{AP} - S_P)/S_{AP}$, with S_{AP} and S_P the corresponding thermopowers in the high- and low-resistance states, respectively, for the FM/Cu NW networks.

Figure 7a–c shows the temperature dependence of the MR and MTP ratios for the Co/Cu (a), Co$_{50}$Ni$_{50}$/Cu (b) and Ni$_{80}$Fe$_{20}$/Cu (c) NW networks. As seen, in all samples, the MR ratio shows a monotonic increase before reaching a plateau at low temperatures. This is expected because of the saturation of the resistivity at low temperatures and the vanishing of the spin mixing effect. Besides, for all three samples, the value of $-$MTP shows a similar increase with decreasing temperature as the MR ratio in the temperature range near RT. In contrast, in the low temperature range, the MTP exhibits very different behaviors depending on the material considered. For the Co/Cu sample, the MTP exhibits a less pronounced effect compared to the MR ratio, which is consistent with previous temperature dependency measurements of the MTP for Co/Cu multilayered NW networks [64]. For the $Co_{50}Ni_{50}$/Cu sample, similar MR and $-$MTP values are observed over the whole investigated temperature range, while for the $Ni_{80}Fe_{20}$/Cu sample, the MTP exhibits a pronounced reinforcement compare to the MR in the low temperature range. Figure 7d shows the ratio between the MR and MTP of the different FM/Cu NW networks, with FM = Co, $Co_{50}Ni_{50}$, $Ni_{80}Fe_{20}$ and Ni, for temperatures in the range of 50 K to 320 K. While the ratios for Co/Cu, $Co_{50}Ni_{50}$/Cu and $Ni_{80}Fe_{20}$/Cu remain below -1.5, the MTP/MR values of the Ni/Cu sample are encompassed between -4 and -8 in the whole temperature range, revealing an enhanced MTP effect with respect to the corresponding MR effect in Ni/Cu multilayered NW networks. A MTP ratio of about -14% was obtained at $T = 80$ K for a corresponding MR slightly below 2% for the Ni/Cu NW network. Such larger value of the MTP with respect to the MR ratio is consistent with our results in Ni homogeneous NW networks, and previous studies in Ni/Cu multilayers [62], and may be ascribed to Seebeck coefficients for spin up and spin down electrons of opposite sign. Furthermore, the magneto-power factor (MPF = $(PF_P - PF_{AP})/PF_{AP}$) can be expressed as MPF = $(1 - MTP)^2/(1 - MR) - 1$. This yields RT MPF ratio of 111%, 155%, 92% and 8% for the Co/Cu, $Co_{50}Ni_{50}$/Cu, $Ni_{80}Fe_{20}$/Cu and Ni/Cu NW networks, respectively.

Table 2. Room-temperature Seebeck coefficient S_P, resistivity ρ_P, power factor PF_P and figure of merit ZT_P obtained in the saturated state for interconnected multilayered nanowire networks made of a stack of successive ferromagnetic metal and Cu layers, as well as their magnetoresistance ratio MR and magneto-thermopower ratio MTP.

	S_P (µV/K)	ρ_P (µΩcm)	PF_P (mW/K^{2}m)	$(ZT)_P$	MR (%)	MTP (%)
Co/Cu	-19.9	8.7	4.6	$1.6 \cdot 10^{-2}$	24.7	-25.1
$Co_{50}Ni_{50}$/Cu	-21.6	10.2	4.6	$1.9 \cdot 10^{-2}$	30.2	-33.7
$Ni_{80}Fe_{20}$/Cu	-24.8	15.3	4.0	$2.5 \cdot 10^{-2}$	17.1	-25.8
Ni/Cu	-10.2	19.8	0.5	$0.4 \cdot 10^{-2}$	0.8	-3.7

Figure 8a–c present evidences that the thermopower is dominated by electron diffusion over the whole temperature range investigated for the different multilayered NW networks considered. Defining the diffusion thermopower $S(H) = eL_0T\rho'(H)/\rho(H)$ by Mott's formula with $\rho'(H) = (d\rho(H)/d\epsilon)_{\epsilon=\epsilon_F}$ the derivative of the electrical resistivity with respect to the energy evaluated at the Fermi level ϵ_F, the diffusion thermopower for antiparallel (AP) and parallel (P) arrangement of the successive FM layer magnetization can be written as $S_{AP} = eL_0T\rho'_{AP}/\rho_{AP}$ and $S_P = eL_0T\rho'_P/\rho_P$. Then, the following expression describing an inverse relationship between the field-dependent thermopower $S(H)$ and electrical resistance $R(H)$ can be obtained [21,57,65]:

$$S(H) = A + \frac{B}{R(H)}, \tag{4}$$

where $A = (S_{AP}R_{AP} - S_PR_P)/(R_{AP} - R_P)$ and $B = R_{AP}R_P(S_P - S_{AP})/(R_{AP} - R_P)$. This expression corresponds to an equivalent form of the Gorter-Nordheim relation for diffusion thermopower in metals and alloys [11], and has been observed at different temperatures in the FM/Cu NW network. This is illustrated by Figures 8a–c, which display $\Delta S(H) = S(H) - S_{AP}$ with respect to $\Delta(1/R(H)) = 1/R(H) - 1/R_{AP}$ for the (a) Co/Cu, (b) $Co_{50}Ni_{50}$/Cu and (c) $Ni_{80}Fe_{20}$/Cu samples. The dashed lines correspond to the theoretical linear relation $\Delta S(H) = B\Delta(1/R(H))$. The data show relatively good accordance with the theoretical linear variation, despite some slight deviation, in particular in the intermediate temperature range.

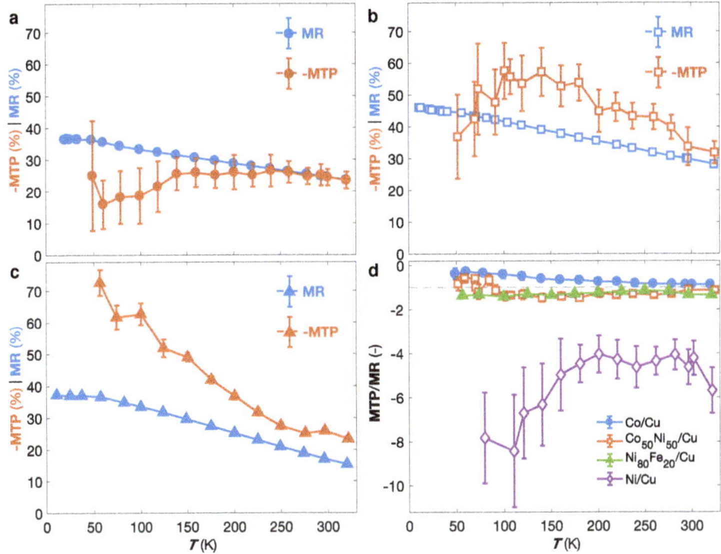

Figure 7. (**a–c**) MR and −MTP values as a function of temperature with the field applied in the plane of the (**a**) Co/Cu, (**b**) Co$_{50}$Ni$_{50}$/Cu and (**c**) Ni$_{80}$Fe$_{20}$/Cu nanowire network films. (**d**) Temperature dependencies of the ratio MTP/MR obtained in IP for the samples in (**a–c**) compared to the Ni/Cu nanowire network. The error bars in (**a–d**) reflect the uncertainty of the electrical and temperature measurements and is set to two times the standard deviation, gathering 95% of the data variation.

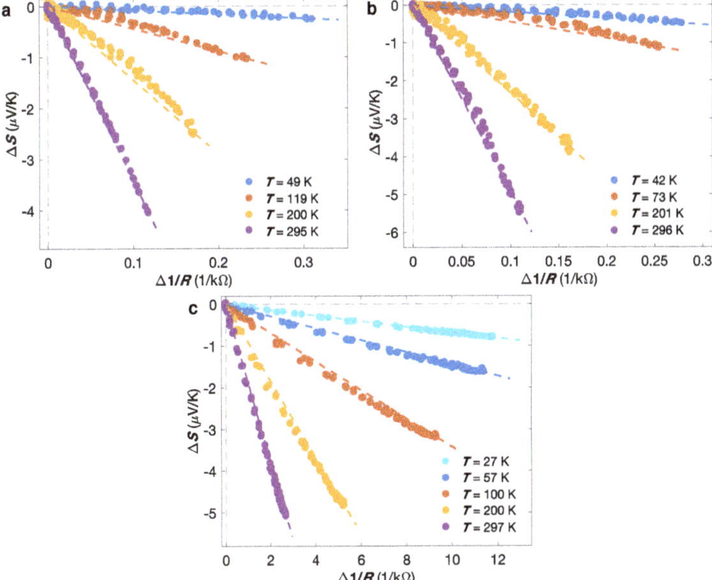

Figure 8. Linear variation of $\Delta S(H) = S(H) - S_{AP}$ vs. $\Delta(1/R(H)) = 1/R(H) - 1/R_{AP}$ at different measured temperatures, illustrating the Gorter-Nordheim characteristics for the (**a**) Co/Cu, (**b**) Co$_{50}$Ni$_{50}$/Cu and (**c**) Ni$_{80}$Fe$_{20}$/Cu nanowire networks. The solid lines correspond to the theoretical relation shown in Equation (4).

In the limits of no-spin relaxation, most of the CPP-GMR data can be understood using a simple two-current series-resistor model, in which the resistance of layers and interfaces simply add and where 'up' and 'down' charge carriers are propagating independently in two spin channels with large spin asymmetries of the electron's scattering [66,67]. Similarly, significantly different Seebeck coefficients for spin-up and spin-down electrons, $S_\uparrow$ and $S_\downarrow$, are expected because the d-band is exchange-split in these ferromagnets, as suggested from previous studies performed on dilute magnetic alloys [46,55]. Assuming that the layers of the magnetic multilayers are thin compared to the spin-diffusion lengths and according to the usual rule when the currents split to flow along two parallel paths (See Figure 9a), the corresponding thermopowers S_{AP} and S_P are simply given by [57]

$$S_{AP} = \frac{S_\uparrow \rho_\uparrow + S_\downarrow \rho_\downarrow}{\rho_\uparrow + \rho_\downarrow} \tag{5}$$

and

$$S_P = \frac{S_\uparrow \rho_\downarrow + S_\downarrow \rho_\uparrow}{\rho_\uparrow + \rho_\downarrow}, \tag{6}$$

where separate resistivities $\rho_\uparrow$ and $\rho_\downarrow$ and Seebeck coefficients $S_\uparrow$ and $S_\downarrow$ are defined for majority and minority spin channels. From Equations (5) and (6), the following relations can be extracted:

$$S_{AP} + S_P = S_\uparrow + S_\downarrow \tag{7}$$

and

$$S_{AP} - S_P = \beta(S_\downarrow - S_\uparrow), \tag{8}$$

where $\beta = (\rho_\downarrow - \rho_\uparrow)/(\rho_\downarrow + \rho_\uparrow)$ denotes the spin asymmetry coefficient for resistivity. Combining Equations (7) and (8), the spin-dependent Seebeck coefficients, $S_\uparrow$ and $S_\downarrow$ can be expressed as follows [13]:

$$S_\uparrow = \frac{1}{2}\left[S_{AP}(1 - \beta^{-1}) + S_P(1 + \beta^{-1})\right], \tag{9}$$

$$S_\downarrow = \frac{1}{2}\left[S_{AP}(1 + \beta^{-1}) + S_P(1 - \beta^{-1})\right]. \tag{10}$$

From Equations (9) and (10), it can be easily deduced that $S_\uparrow = S_P$ and $S_\downarrow = S_{AP}$ in the limits of an extremely large MR ratio ($\beta \to 1$).

The temperature dependencies of S_{AP}, S_P, $S_\uparrow$ and $S_\downarrow$ are shown in Figures 9b–d for the Co/Cu, $Co_{50}Ni_{50}$/Cu and $Ni_{80}Fe_{20}$/Cu multilayered NW networks. Below RT, the various Seebeck coefficients of all three samples decrease almost linearly with decreasing temperature, which is another indicative of the dominance of diffusion thermopower. The deviation from the linear regime in Figure 9b–d can be ascribed to other possible contribution to the thermopower such as magnon-drag. The estimated values at RT of $S_\uparrow$ and $S_\downarrow$ are reported in Table 3, using $\beta = MR^{1/2}$ in Equations (9) and (10). The values of $S_\uparrow$ and $S_\downarrow$ for Co/Cu NWs are similar to those previously reported in bulk Co ($S_\uparrow = -30$ µV/K and $S_\downarrow = -12$ µV/K) [46]. In contrast, the value for $\Delta S = S_\uparrow - S_\downarrow$ for Co/Cu, $Co_{50}Ni_{50}$/Cu and $Ni_{80}Fe_{20}$/Cu NWs shown in Table 3, in the range of -8 to -12 µV/K [13,14,22], are much larger than the ones of -1.8 µV/K and -3.8 µV/K to -4.5 µV/K extracted from measurements performed on Co/Cu/Co nanopillar spin valve and $Ni_{80}Fe_{20}$/Cu/$Ni_{80}Fe_{20}$ nanopillar and lateral spin devices, respectively, using a 3D finite-element model [6,68]. The largest spin-dependent Seebeck coefficient is found in $Ni_{80}Fe_{20}$/Cu NWs, reaching -12.3 µV/K [22]. From the estimated values of $S_\uparrow$ and $S_\downarrow$, the RT spin asymmetry for Seebeck coefficients $\eta = (S_\downarrow - S_\uparrow)/(S_\downarrow + S_\uparrow)$ has been estimated, and are reported in Table 3 for the interconnected Co/Cu, $Co_{50}Ni_{50}$/Cu and $Ni_{80}Fe_{20}$/Cu NWs, respectively. It is also found that the η value found for the crossed Co/Cu NW network is consistent with the values previously reported for parallel Co/Cu NWs [64].

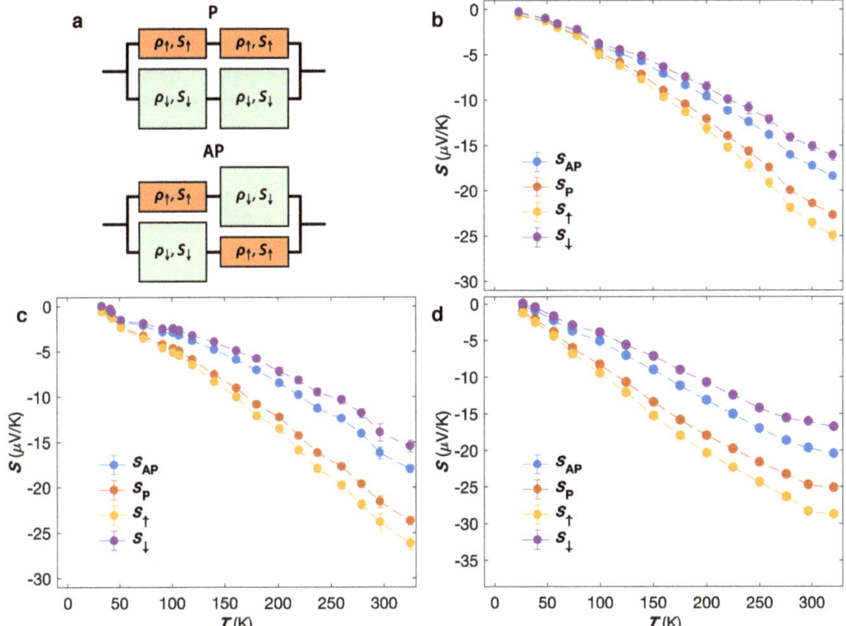

Figure 9. (**a**) The two-current model for the resistivity and the thermopower considering both parallel (P) and antiparallel (AP) magnetic configurations. (**b**–**d**) Measured Seebeck coefficients at zero applied fields S_{AP} (blue circles) and at saturating magnetic fields S_P (red circles) of interconnected (**b**) Co/Cu, (**c**) $Co_{50}Ni_{50}$/Cu and (**d**) $Ni_{80}Fe_{20}$/Cu nanowire networks, along with the corresponding calculated spin-dependent Seebeck coefficients $S_\uparrow$ (orange circles) and $S_\downarrow$ (violet circles) using Equations (9) and (10). The error bars reflect the uncertainty of the electrical and temperature measurements and is set to two times the standard deviation, gathering 95% of the data variation.

Table 3. Room-temperature spin-dependent Seebeck coefficients, $S_\uparrow$ and $S_\downarrow$, of the FM/Cu nanowire networks with FM = Co, $Co_{50}Ni_{50}$, $Ni_{80}Fe_{20}$ an Ni, along with $\Delta S = S_\uparrow - S_\downarrow$ and the spin asymmetry coefficients for resistivity β and Seebeck coefficient η.

	$S_\uparrow$ (μV/K)	$S_\downarrow$ (μV/K)	ΔS (μV/K)	β (-)	η (-)
Co/Cu	−23.6	−15.1	−8.5	0.50	−0.22
$Co_{50}Ni_{50}$/Cu	−23.9	−13.9	−10.0	0.55	−0.26
$Ni_{80}Fe_{20}$/Cu	−28.4	−16.1	−12.3	0.41	−0.28

In the limit of MR = β^2, the magnetothermopower can also be expressed as [13]:

$$\text{MTP} = \frac{2\beta\eta}{1+\beta\eta}. \tag{11}$$

Considering first the case $|\beta\eta| \ll 1$, this leads to MTP/MR $\approx 2\eta/\beta$, which means that enhancement of the MTP ratio compared to the corresponding MR ratio is expected if $2|\eta| > |\beta|$. The amplitudes of the MR and MTP effects for the Co/Cu, $Co_{50}Ni_{50}$/Cu and $Ni_{80}Fe_{20}$/Cu NW networks studied have been found comparable at RT, with $2|\eta| \approx |\beta|$ as seen in Table 3. Similarities between the amplitudes of the MR and MTP at RT were already observed in arrays of parallel Co/Cu NWs [64,69,69] and Co/Cu CIP multilayers [57,70]. However, the physical reason of this is unclear. In contrast to the other FM/Cu NW networks, the Ni/Cu system shows a much larger MTP ratio than its corresponding MR ratio, as seen in Figure 7d. This significant boost of the MTP, compared to MR may indicate an underlying large η coefficient due to contrasting $S_\uparrow$ and $S_\downarrow$ values, while β is expected to be small in Ni. This is consistent with the values of $S_\uparrow = -43$ µV/K and $S_\downarrow = 20$ µV/K reported by Cadeville and Roussell [46], leading to $\eta \approx -3$. Furthermore, from Equation (11), infinitely large MTP effect is expected when the product $\beta\eta$ tends to -1. While $|\beta| < 1$, $|\eta| > 1$ can be reached if $S_\uparrow$ and $S_\downarrow$ exhibit opposite signs. Therefore, the fabrication of multilayered NWs with appropriate magnetic layer composition should make it possible to fine-tune the power factor of thermoelectric energy conversion with an external magnetic field. The limit case of an infinite MTP ratio underlies that the Seebeck coefficient obtained in the antiparallel state S_{AP} reaches 0. Indeed, the Seebeck coefficients in the antiparallel and parallel states can be expressed as

$$S_{AP} = \frac{S_\uparrow + S_\downarrow}{2}(1 + \beta\eta) \tag{12}$$

and

$$S_P = \frac{S_\uparrow + S_\downarrow}{2}(1 - \beta\eta). \tag{13}$$

For $\beta\eta = -1$, this yields $S_{AP} = 0$ and $S_P = S_\uparrow + S_\downarrow$. As a consequence, for a practice device, Seebeck coefficients for spin up and spin down electrons of opposite sign and with largely asymmetrical amplitudes are required. This would lead to a system where at zero magnetic field, no thermoelectric current is generated, while under an applied external magnetic field, a large thermoelectric current is generated. It should be noted that the limit case $\beta\eta = 1$ yields the opposite scenario where $S_{AP} = S_\uparrow + S_\downarrow$ and $S_P = 0$. In that case, a thermoelectric current is generated in absence of magnetic fields, which vanishes under an applied external magnetic field. The case $\beta\eta \rightarrow 1$ also requires $|\eta| > 1$ and therefore Seebeck coefficients for spin up and spin down with opposite signs. Moreover, Equation (7) indicates that the more different their amplitudes, the larger the S_{AP} value, which is required for practical applications. In this context, the Ni/Cu system is expected to be a good potential candidate for the observation of highly enhanced MTP effect, since $\beta > 0$ and, according to Cadeville and Roussel [46], $\eta \approx -3$. Experimental challenge lies with making interconnected Ni/Cu NW networks with larger magneto-transport properties.

Interestingly, it can be shown that the diffusion thermopower of a ferromagnetic homogeneous NW network can be expressed as

$$S_{FM} = \frac{S_\uparrow \alpha + S_\downarrow}{\alpha + 1}. \tag{14}$$

Therefore, because $\alpha \gg 1$ is expected for Co, the Seebeck coefficient obtained for homogeneous Co NW networks (-28 µV/K, see Table 1) is consistent with $S_{Co} \approx S_{Co,\uparrow}$. Moreover, the addition of a very small amount of Cr in Co is expected to induce $\alpha \ll 1$. As a consequence, the Seebeck coefficient of dilute CoCr alloys can be approximated by $S_{Co_{99}Cr_1} \approx S_{Co,\downarrow}$, assuming that the very small amount of Cr does not significantly impact the Seebeck coefficient of spin down electron, which is also in good agreement with the value reported for $Co_{99}Cr_1$ (-12 µV/K). In contrast, because $\alpha \approx 1$ is expected for Ni, Equation (14) leads to $S_{Ni} = (S_\uparrow + S_\downarrow)/2$. The values of $S_{Ni} \approx -20$ µV/K is found in good agreement with the Seebeck coefficients for spin up and spin down electrons reported by Cadeville and Roussell ($S_\uparrow \approx -43$ µV/K and $S_\downarrow \approx +20$µV/K [46]). Moreover, the addition of diluted Cr impurities in Ni is also expected to induce $\alpha \ll 1$. Therefore, the Seebeck coefficient of dilute NiCr alloys is expected to tend towards the value of $S_{Ni,\downarrow}$. The positive values obtained in dilute NiCr alloys is in good agreement with the positive value of $S_{Ni,\downarrow}$ reported by Cadeville and Roussell [46] as shown in Figure 4c. This is another indication that Ni may exhibit a probable spin-dependent Seebeck coefficients of opposite sign.

4. Conclusions

This research provides a simple and cost-effective pathway to fabricate highly efficient and large-scale nanowire-based thermoelectric films meeting key requirements for electrical, thermal and mechanical stability. Flexible and planar interconnected nanowire networks allows for both p- and n-type thermoelectric modules with large room-temperature Seebeck coefficient and power factors.

Since there is no sample size limitation, this fabrication method is expandable into network films with much larger dimensions. A practical planar thermoelectric cooler made of flexible and shapeable thermoelectric modules consisting of stacked nanowire network films that are connected electrically in series and thermally in parallel can be easily obtained. Furthermore, an unexpected high value of the magneto-thermopower effect compared with that of the corresponding magnetoresistance effect has been observed in Ni crossed nanowires, in agreement with previous studies on Ni nanowires.

Embedded nanowire networks in porous polymer films are also perspective materials for spin caloritronics applications. Centimetre-scale interconnected network films made of multilayered nanowires show giant magnetoresistance and giant magneto-thermoelectric effects together with large and magnetically modulated room-temperature thermoelectric power factor up to 5 mW/K^2m. These macroscopic nanowire networks also enable the direct extraction of key material parameters for spin caloritronics such as spin-dependent Seebeck coefficients. Spin-dependent Seebeck coefficients up to -12.3 µV/K have been obtained at room temperature in $Ni_{80}Fe_{20}$/Cu multilayered nanowire networks. Moreover, a large enhancement of the magneto-thermopower ratio compared to the corresponding magnetoresistance ratio of about four times larger at room temperature has been measured in Ni/Cu nanowire networks, potentially indicating high spin-dependent Seebeck coefficients in Ni, consistent with previous studies. These results open an exciting and a promising pathway for the next generation of flexible and lightweight thermoelectric devices exploiting the spin degree of freedom and the realization of magnetic thermal switch for heat management. In addition, flexible thermoelectric films based on macroscopic networks of interconnected nanowires can also be used in applications for devices with low energy requirements where the heat input is essentially free, as in the case of waste heat.

Author Contributions: Conceptualization, L.P.; Funding acquisition, L.P.; Investigation, T.d.C.S.C.G. and N.M.; Project administration, L.P.; Software, F.A.A.; Supervision, L.P.; Validation, T.d.C.S.C.G., N.M., F.A.A. and L.P.; Writing—original draft, T.d.C.S.C.G.; Writing—review & editing, T.d.C.S.C.G., N.M., F.A.A. and L.P. All authors have read and agreed to the published version of the manuscript.

Funding: Financial support was provided by Wallonia/Brussels Community (ARC 18/23-093) and the Belgian Fund for Scientific Research (FNRS).

Acknowledgments: T.d.C.S.C.G. is a Research Fellow of the FNRS. N.M. acknowledges the Research Science Foundation of Belgium (FRS-FNRS) for financial support (FRIA grant). F.A.A. is a Postdoctoral Researcher of the FNRS. The authors would like to thank E. Ferain and the it4ip Company for supplying polycarbonate membranes.

Conflicts of Interest: The authors declare no conflict of interest.

References

1. He, J.; Tritt, T.M. Advances in thermoelectric materials research: Looking back and moving forward. *Science* **2017**, *357*. [CrossRef] [PubMed]
2. Bauer, G.E.W.; Saitoh, E.; van Wees, B.J. Spin caloritronics. *Nat. Mater.* **2012**, *11*, 391. [CrossRef] [PubMed]
3. Boona, S.R.; Myers, R.C.; Heremans, J.P. Spin caloritronics. *Energy Environ. Sci.* **2014**, *7*, 885–910. [CrossRef]
4. Uchida, K.; Xiao, J.; Adachi, H.; Ohe, J.; Takahashi, S.; Ieda, J.; Ota, T.; Kajiwara, Y.; Umezawa, H.; Kawai, H.; et al. Spin Seebeck insulator. *Nat. Mater.* **2010**, *9*, 894. [CrossRef] [PubMed]
5. Jaworski, C.M.; Yang, J.; Mack, S.; Awschalom, D.D.; Heremans, J.P.; Myers, R.C. Observation of the spin-Seebeck effect in a ferromagnetic semiconductor. *Nat. Mater.* **2010**, *9*, 898. [CrossRef] [PubMed]
6. Slachter, A.; Bakker, F.L.; Adam, J.P.; van Wees, B.J. Thermally driven spin injection from a ferromagnet into a non-magnetic metal. *Nat. Phys.* **2010**, *6*, 879. [CrossRef]
7. Hatami, M.; Bauer, G.E.W.; Zhang, Q.; Kelly, P.J. Thermal Spin-Transfer Torque in Magnetoelectronic Devices. *Phys. Rev. Lett.* **2007**, *99*, 066603. [CrossRef]
8. Pushp, A.; Phung, T.; Rettner, C.; Hughes, B.P.; Yang, S.H.; Parkin, S.S.P. Giant thermal spin-torque–assisted magnetic tunnel junction switching. *Proc. Natl. Acad. Sci. USA* **2015**, *112*, 6585–6590. [CrossRef]
9. Liebing, N.; Serrano-Guisan, S.; Rott, K.; Reiss, G.; Langer, J.; Ocker, B.; Schumacher, H.W. Tunneling Magnetothermopower in Magnetic Tunnel Junction Nanopillars. *Phys. Rev. Lett.* **2011**, *107*, 177201. [CrossRef]

10. Walter, M.; Walowski, J.; Zbarsky, V.; Münzenberg, M.; Schäfers, M.; Ebke, D.; Reiss, G.; Thomas, A.; Peretzki, P.; Seibt, M.; et al. Seebeck effect in magnetic tunnel junctions. *Nat. Mater.* **2011**, *10*, 742–746. [CrossRef]

11. Blatt, F.J.; Schroeder, P.A.; Foiles, C.L.; Greig, D. *Thermoelectric Power of Metals*; Springer: Boston, MA, USA, 1976.

12. Watzman, S.J.; Duine, R.A.; Tserkovnyak, Y.; Boona, S.R.; Jin, H.; Prakash, A.; Zheng, Y.; Heremans, J.P. Magnon-drag thermopower and Nernst coefficient in Fe, Co, and Ni. *Phys. Rev. B* **2016**, *94*, 144407. [CrossRef]

13. da Câmara Santa Clara Gomes, T.; Abreu Araujo, F.; Piraux, L. Making flexible spin caloritronic devices with interconnected nanowire networks. *Sci. Adv.* **2019**, *5*, eaav2782. [CrossRef]

14. Abreu Araujo, F.; da Câmara Santa Clara Gomes, T.; Piraux, L. Magnetic Control of Flexible Thermoelectric Devices Based on Macroscopic 3D Interconnected Nanowire Networks. *Adv. Electron. Mater.* **2019**, *5*, 1800819. [CrossRef]

15. Rauber, M.; Alber, I.; Müller, S.; Neumann, R.; Picht, O.; Roth, C.; Schökel, A.; Toimil-Molares, M.E.; Ensinger, W. Highly-Ordered Supportless Three-Dimensional Nanowire Networks with Tunable Complexity and Interwire Connectivity for Device Integration. *Nano Lett.* **2011**, *11*, 2304–2310. [CrossRef] [PubMed]

16. Araujo, E.; Encinas, A.; Velázquez-Galván, Y.; Martinez-Huerta, J.M.; Hamoir, G.; Ferain, E.; Piraux, L. Artificially modified magnetic anisotropy in interconnected nanowire networks. *Nanoscale* **2015**, *7*, 1485–1490. [CrossRef] [PubMed]

17. da Câmara Santa Clara Gomes, T.; de la Torre Medina, J.; Velázquez-Galván, Y.G.; Martínez-Huerta, J.M.; Encinas, A.; Piraux, L. Interplay between the magnetic and magneto-transport properties of 3D interconnected nanowire networks. *J. Appl. Phys.* **2016**, *120*, 043904. [CrossRef]

18. da Câmara Santa Clara Gomes, T.; De La Torre Medina, J.; Lemaitre, M.; Piraux, L. Magnetic and Magnetoresistive Properties of 3D Interconnected NiCo Nanowire Networks. *Nanoscale Res. Lett.* **2016**, *11*, 466. [CrossRef]

19. da Câmara Santa Clara Gomes, T.; Medina, J.D.L.T.; Velázquez-Galván, Y.G.; Martínez-Huerta, J.M.; Encinas, A.; Piraux, L. 3-D Interconnected Magnetic Nanofiber Networks With Multifunctional Properties. *IEEE Trans. Magn.* **2017**, *53*, 1–6. [CrossRef]

20. da Câmara Santa Clara Gomes, T.; Marchal, N.; Abreu Araujo, F.; Piraux, L. Tunable magnetoresistance and thermopower in interconnected NiCr and CoCr nanowire networks. *Appl. Phys. Lett.* **2019**, *115*, 242402. [CrossRef]

21. Piraux, L.; George, J.M.; Despres, J.F.; Leroy, C.; Ferain, E.; Legras, R.; Ounadjela, K.; Fert, A. Giant magnetoresistance in magnetic multilayered nanowires. *Appl. Phys. Lett.* **1994**, *65*, 2484–2486. [CrossRef]

22. Marchal, N.; da Câmara Santa Clara Gomes, T.; Abreu Araujo, F.; Piraux, L. Large Spin-Dependent Thermoelectric Effects in NiFe-based Interconnected Nanowire Networks. *Nanoscale Res. Lett.* **2020**, *15*, 137. [CrossRef] [PubMed]

23. Rowe, D.M. *CRC Handbook of Thermoelectrics*; CRC Press: Boca Raton, FL, USA, 1995.

24. Ho, C.Y.; Bogaard, R.H.; Chi, T.C.; Havill, T.N.; James, H.M. Thermoelectric power of selected metals and binary alloy systems. *Thermochim. Acta* **1993**, *218*, 29–56. [CrossRef]

25. Basargin, O.V.; Zarkhov, A. Thermal E.M.F. of iron-nickel alloys with an F.C.C. lattice. *Fiz. Metal. Met.* **1974**, *37*, 891–894.

26. Tanji, Y.; Moriya, H.; Nakagawa, Y. Anomalous Concentration Dependence of Thermoelectric Power of Fe-Ni (fcc) Alloys at High Temperatures. *J. Phys. Soc. Jpn.* **1978**, *45*, 1244–1248. [CrossRef]

27. Böhnert, T.; Vega, V.; Michel, A.K.; Prida, V.M.; Nielsch, K. Magneto-thermopower and magnetoresistance of single Co-Ni alloy nanowires. *Appl. Phys. Lett.* **2013**, *103*, 092407. [CrossRef]

28. Niemann, A.C.; Böhnert, T.; Michel, A.K.; Bäßler, S.; Gotsmann, B.; Neuróhr, K.; Tóth, B.; Péter, L.; Bakonyi, I.; Vega, V.; et al. Thermoelectric Power Factor Enhancement by Spin-Polarized Currents—A Nanowire Case Study. *Adv. Electron. Mater.* **2016**, *2*, 1600058. [CrossRef]

29. Kamalakar, M.V.; Raychaudhuri, A.K. Low temperature electrical transport in ferromagnetic Ni nanowires. *Phys. Rev. B* **2009**, *79*, 205417. [CrossRef]

30. Meaden, G.T. *Electrical Resistance of Metals*; Springer: Berlin/Heidelberg, Germany, 1965.

31. Tóth, B.G.; Péter, L.; Révész, Á.; Pádár, J.; Bakonyi, I. Temperature dependence of the electrical resistivity and the anisotropic magnetoresistance (AMR) of electrodeposited Ni-Co alloys. *Eur. Phys. J. B* **2010**, *75*, 167–177. [CrossRef]

32. Ho, C.Y.; Ackerman, M.W.; Wu, K.Y.; Havill, T.N.; Bogaerd, R.H.; Matula, R.A.; Ob, S.G.; James, H.M. *Electrical Resistivity of Selected Binary Alloy Systems*; INDAS Report 59; NBS: Washington, DC, USA, 1981.

33. Yao, Y.D.; Arajs, S.; Anderson, E.E. Electrical resistivity of nickel-rich nickel-chromium alloys between 4 and 300 K. *J. Low Temp. Phys.* **1975**, *21*, 369–376. [CrossRef]

34. Bubnova, O.; Khan, Z.U.; Malti, A.; Braun, S.; Fahlman, M.; Berggren, M.; Crispin, X. Optimization of the thermoelectric figure of merit in the conducting polymer poly(3,4-ethylenedioxythiophene). *Nat. Mater.* **2011**, *10*, 429. [CrossRef]

35. Du, Y.; Xu, J.; Paul, B.; Eklund, P. Flexible thermoelectric materials and devices. *Appl. Mater. Today* **2018**, *12*, 366–388. [CrossRef]

36. Vandaele, K.; Watzman, S.J.; Flebus, B.; Prakash, A.; Zheng, Y.; Boona, S.R.; Heremans, J.P. Thermal spin transport and energy conversion. *Mater. Today Phys.* **2017**, *1*, 39–49. [CrossRef]

37. Boona, S.R.; Morelli, D.T. Enhanced thermoelectric properties of $CePd_{3-x}Pt_x$. *Appl. Phys. Lett.* **2012**, *101*, 101909. [CrossRef]

38. Adams, M.; Verosky, M.; Zebarjadi, M.; Heremans, J. Active Peltier Coolers Based on Correlated and Magnon-Drag Metals. *Phys. Rev. Appl.* **2019**, *11*, 054008. [CrossRef]

39. Zebarjadi, M. Electronic cooling using thermoelectric devices. *Appl. Phys. Lett.* **2015**, *106*, 203506. [CrossRef]

40. Ou, M.N.; Yang, T.J.; Harutyunyan, S.R.; Chen, Y.Y.; Chen, C.D.; Lai, S.J. Electrical and thermal transport in single nickel nanowire. *Appl. Phys. Lett.* **2008**, *92*, 063101. [CrossRef]

41. Ho, C.; Chi, T.; Bogaard, R.; Havill, T.; James, H. Thermoelectric power of selected metals and binary alloy systems. In *Thermal Conductivity 17, Proceedings of the 17th International Thermal Conductivity Conference, Gaithersburg, Maryland*; Plenum Press: New York, NY, USA, 1981; pp. 195–205.

42. Smit, J. Magnetoresistance of ferromagnetic metals and alloys at low temperatures. *Physica* **1951**, *17*, 612–627. [CrossRef]

43. McGuire, T.; Potter, R. Anisotropic magnetoresistance in ferromagnetic 3d alloys. *IEEE Trans. Magn.* **1975**, *11*, 1018–1038. [CrossRef]

44. Avery, A.D.; Pufall, M.R.; Zink, B.L. Determining the planar Nernst effect from magnetic-field-dependent thermopower and resistance in nickel and permalloy thin films. *Phys. Rev. B* **2012**, *86*, 184408. [CrossRef]

45. Mitdank, R.; Handwerg, M.; Steinweg, C.; Töllner, W.; Daub, M.; Nielsch, K.; Fischer, S.F. Enhanced magneto-thermoelectric power factor of a 70 nm Ni-nanowire. *J. Appl. Phys.* **2012**, *111*, 104320. [CrossRef]

46. Cadeville, M.C.; Roussel, J. Thermoelectric power and electronic structure of dilute alloys of nickel and cobalt with d transition elements. *J. Phys. F Met. Phys.* **1971**, *1*, 686. [CrossRef]

47. Campbell, I.; Fert, A. Transport properties of ferromagnets. *Handb. Ferromagn. Mater.* **1982**, *3*, 747–804. [CrossRef]

48. Inoue, J.I. CHAPTER 2-GMR, TMR and BMR. In *Nanomagnetism and Spintronics*; Shinjo, T., Ed.; Elsevier: Amsterdam, The Netherlands, 2009; pp. 15–92. [CrossRef]

49. Van Elst, H.C. The anisotropy in the magneto-resistance of some nickel alloys. *Physica* **1959**, *25*, 708–720. [CrossRef]

50. Campbell, I.A.; Fert, A.; Jaoul, O. The spontaneous resistivity anisotropy in Ni-based alloys. *J. Phys. C Solid State Phys.* **1970**, *3*, S95. [CrossRef]

51. Dorleijn, J.W.F.; Miedema, A.R. An investigation of the resistivity anisotropy in nickel alloys. *J. Phys. F Met. Phys.* **1975**, *5*, 1543–1553. [CrossRef]

52. McGuire, T.; Aboaf, J.; Klokholm, E. Negative anisotropic magnetoresistance in 3d metals and alloys containing iridium. *IEEE Trans. Magn.* **1984**, *20*, 972–974. [CrossRef]

53. Jaoul, O.; Campbell, I.; Fert, A. Spontaneous resistivity anisotropy in Ni alloys. *J. Magn. Magn. Mater.* **1977**, *5*, 23–34. [CrossRef]

54. Durand, J.; Gautier, F. Conduction a deux bandes et effet de periode dans les alliages a base de nickel et de cobalt. *J. Phys. Chem. Solids* **1970**, *31*, 2773–2787. [CrossRef]

55. Farrell, T.; Greig, D. The thermoelectric power of nickel and its alloys. *J. Phys. C Solid State Phys.* **1970**, *3*, 138. [CrossRef]

56. MacDonald, D.K.C. *Thermoelectricity: An Introduction to the Principles*; Wiley: Mineola, NY, USA, 1962.

57. Shi, J.; Parkin, S.S.P.; Xing, L.; Salamon, M.B. Giant magnetoresistance and magnetothermopower in Co/Cu multilayers. *J. Magn. Magn. Mater.* **1993**, *125*, L251–L256. [CrossRef]

58. Mayadas, A.F.; Janak, J.F.; Gangulee, A. Resistivity of Permalloy thin films. *J. Appl. Phys.* **1974**, *45*, 2780–2781. [CrossRef]

59. de la Torre Medina, J.; Darques, M.; Encinas, A.; Piraux, L. Dipolar interactions in multilayered Co0.96Cu0.04/Cu nanowire arrays. *Phys. Status Solidi* **2008**, *205*, 1813–1816. [CrossRef]

60. Böhnert, T.; Niemann, A.C.; Michel, A.K.; Bäßler, S.; Gooth, J.; Tóth, B.G.; Neuróhr, K.; Péter, L.; Bakonyi, I.; Vega, V.; et al. Magnetothermopower and magnetoresistance of single Co-Ni/Cu multilayered nanowires. *Phys. Rev. B* **2014**, *90*, 165416. [CrossRef]

61. Sato, H.; Miya, S.; Kobayashi, Y.; Aoki, Y.; Yamamoto, H.; Nakada, M. Magnetoresistance, Hall effect, and thermoelectric power in spin valves. *J. Appl. Phys.* **1998**, *83*, 5927–5932. [CrossRef]

62. Serrano-Guisan, S.; Gravier, L.; Abid, M.; Ansermet, J.P. Thermoelectrical study of ferromagnetic nanowire structures. *J. Appl. Phys.* **2006**, *99*, 08T108. [CrossRef]

63. Cahaya, A.B.; Tretiakov, O.A.; Bauer, G.E.W. Spin Seebeck power generators. *Appl. Phys. Lett.* **2014**, *104*, 042402. [CrossRef]

64. Gravier, L.; Fábián, A.; Rudolf, A.; Cachin, A.; Wegrowe, J.E.; Ansermet, J.P. Spin-dependent thermopower in Co/Cu multilayer nanowires. *J. Magn. Magn. Mater.* **2004**, *271*, 153–158. [CrossRef]

65. Shi, J.; Kita, E.; Xing, L.; Salamon, M.B. Magnetothermopower of a $Ag_{80}Co_{20}$ granular system. *Phys. Rev. B* **1993**, *48*, 16119–16122. [CrossRef]

66. Lee, S.F.; Pratt, W.P.; Loloee, R.; Schroeder, P.A.; Bass, J. "Field-dependent interface resistance" of Ag/Co multilayers. *Phys. Rev. B* **1992**, *46*, 548–551. [CrossRef]

67. Bass, J. CPP magnetoresistance of magnetic multilayers: A critical review. *J. Magn. Magn. Mater.* **2016**, *408*, 244–320. [CrossRef]

68. Dejene, F.K.; Flipse, J.; van Wees, B.J. Spin-dependent Seebeck coefficients of $Ni_{80}Fe_{20}$ and Co in nanopillar spin valves. *Phys. Rev. B* **2012**, *86*, 024436. [CrossRef]

69. Baily, S.A.; Salamon, M.B.; Oepts, W. Magnetothermopower of cobalt/copper multilayers with gradient perpendicular to planes. *J. Appl. Phys.* **2000**, *87*, 4855–4857. [CrossRef]

70. Nishimura, K.; Sakurai, J.; Hasegawa, K.; Saito, Y.; Inomata, K.; Shinjo, T. Thermoelectric Power of Co/Cu Multilayer. *J. Phys. Soc. Jpn.* **1994**, *63*, 2685–2690. [CrossRef]

Publisher's Note: MDPI stays neutral with regard to jurisdictional claims in published maps and institutional affiliations.

 *nanomaterials*

Review

Focused-Electron-Beam Engineering of 3D Magnetic Nanowires

César Magén [1,2,3,*], Javier Pablo-Navarro [1,2,4] and José María De Teresa [1,2,3]

1   Instituto de Nanociencia y Materiales de Aragón (INMA), Universidad de Zaragoza-CSIC,
    50009 Zaragoza, Spain; j.pablo-navarro@hzdr.de (J.P.-N.); deteresa@unizar.es (J.M.D.T.)
2   Laboratorio de Microscopías Avanzadas (LMA), Universidad de Zaragoza, 50018 Zaragoza, Spain
3   Departamento de Física de la Materia Condensada, Universidad de Zaragoza, 50009 Zaragoza, Spain
4   Institute of Ion Beam Physics and Materials Research, Helmholtz-Zentrum Dresden-Rossendorf,
    01328 Dresden, Germany
*   Correspondence: cmagend@unizar.es; Tel.: +34-876-555369; Fax: +34-976-762-776

Abstract: Focused-electron-beam-induced deposition (FEBID) is the ultimate additive nanofabrication technique for the growth of 3D nanostructures. In the field of nanomagnetism and its technological applications, FEBID could be a viable solution to produce future high-density, low-power, fast nanoelectronic devices based on the domain wall conduit in 3D nanomagnets. While FEBID has demonstrated the flexibility to produce 3D nanostructures with almost any shape and geometry, the basic physical properties of these out-of-plane deposits are often seriously degraded from their bulk counterparts due to the presence of contaminants. This work reviews the experimental efforts to understand and control the physical processes involved in 3D FEBID growth of nanomagnets. Co and Fe FEBID straight vertical nanowires have been used as benchmark geometry to tailor their dimensions, microstructure, composition and magnetism by smartly tuning the growth parameters, post-growth purification treatments and heterostructuring.

Keywords: nanomagnetism; focused-electron-beam-induced deposition; nanofabrication; nanolithography; magnetic nanowires; three-dimensional; core-shell; purification; thermal annealing; electron holography

Citation: Magén, C.; Pablo-Navarro, J.; De Teresa, J.M. Focused-Electron-Beam Engineering of 3D Magnetic Nanowires. *Nanomaterials* **2021**, *11*, 2. https://doi.org/10.3390/nano11020402

Received: 16 December 2020
Accepted: 30 January 2021
Published: 4 February 2021

Publisher's Note: MDPI stays neutral with regard to jurisdictional claims in published maps and institutional affiliations.

1. Introduction

Three-dimensional (3D) magnetic nanostructures are the playground of a wide range of exciting physical phenomena in nanomagnetism, where curved geometries enable the onset of new types of exotic magnetic configurations, such as topologically protected and chiral magnetic textures [1–4]. Furthermore, 3D nanomagnets present several key features to postulate a paradigmatic solution to different challenges that the semiconductor industry must confront in the years to come to reconcile continued miniaturization, increasing performance and reduced power consumption [5–7]. The transition to 3D nanoarchitectures would overcome the intrinsic areal density limitations of conventional CMOS technologies caused by increasing leakage currents due to quantum effects [8]. They may also reduce the power consumption by storing and processing memory by ultrafast magnetic domain walls or skyrmions driven by low-power spin currents [9–11].

Even though the growth of high-purity, narrow 3D ferromagnetic structures may be the gateway to a broad range of opportunities for both fundamental and technological applications, most standard nanolithography techniques lack the flexibility to implement complex 3D structures at the nanoscale. For this purpose, additive manufacturing approaches to nanofabrication present an ideal solution. Among them, focused-electron-beam-induced deposition (FEBID) presents unique properties to become the ultimate 3D nano-printing technique [12]. FEBID is a single-step additive nanolithography technique based on the local decomposition of the molecules of an organometallic precursor gas, adsorbed on the surface of a substrate, thus producing a solid deposit [13,14]. This mechanism is mainly driven by the interaction of the primary beam and the secondary electrons

emitted by the substrate with the adsorbed precursor molecules [15]. This basic principle confers a great control on the material deposit, geometry, and growth conditions t
design conducting, insulating, superconducting, plasmonic or ferromagnetic structure
with virtually any shape and nanometer resolution, in 2D [16] and 3D [17]. Over th
years, numerous types of FEBID deposits have been developed as structural materials [18
for electrical contacting [19], nanosensing [20] or plasmonic structures [21]. In particula
the growth of 3D ferromagnets by FEBID has been fruitful and yielded highly sophisti
cated architectures. Remarkable applications have been developed in magnetic sensing b
functionalization of magnetic probes for magnetic force microscopy [22] and ferromagnet
resonance force microscopy [23] in materials science and biology [24], as magneticall
driven mechanical nano-actuators [25], 3D domain wall conduit [26], ferromagnetic de
signs based on 3D FEBID scaffolds [18], arrays of 3D nanopillars for magnetic logic [27
3D artificial ferromagnetic lattices [28,29], and magnetically chiral 3D architectures [30
Further examples of applications of 3D FEBID ferromagnets have been reviewed recentl
by Fernández-Pacheco et al. [31].

One of the main difficulties of FEBID growth for the design of high-performanc
nanodevices is to obtain the desired geometry and dimensions with a high level of purit
The presence of a degree of contaminants in the deposit is inherent to the FEBID proces
as some of the precursor residues are easily integrated into the deposit together with th
active material [32], while secondary electrons emitted away from beam position induc
the formation of an extended halo [33]. These drawbacks have motivated a dedicate
effort for optimization of FEBID growth conditions, beginning with the choice of th
gas precursor. Numerous precursor molecules have been explored for the growth o
ferromagnetic materials. A detailed account of the precursors reported in the literatur
for 2D growth of ferromagnets is beyond the scope of this review, and has been reporte
elsewhere [34], but it is worth discussing some key aspects. Most of the precursors that hav
been reported are organometallic complexes based on carbonyl (CO) groups, thus C and
are the expected impurities derived from incomplete precursor decomposition. The mo
widely used is dicobalt octacarbonyl, $Co_2(CO)_8$, for which 2D deposits with 95 at. % C
content with metallic conduction have been achieved without further post-processing [35
even in halo-free extremely narrow (<30 nm) nanowires [36]. $Co(CO)_3NO$ has also bee
tested for 2D FEBID growth, and metal contents of 50–55 at. % have been obtained [37,38
In the case of iron, the main precursors used are $Fe(CO)_5$ [39] and $Fe_2(CO)_9$ [40], and met
contents above 75 at. % have been obtained. Iron precursors evidence the importance o
residual gases in the deposition chamber, and 95 at. % purity can be achieved in ultra-hig
vacuum conditions with $Fe(CO)_5$ [41]. The use of heteronuclear precursors allows one t
grow alloyed magnetic materials; for instance, the precursor $HCo_3Fe(CO)_{12}$ has been use
to produce Co_3Fe deposits with metallic contents as high as 80 at. % [42], and has also bee
grown in 3D [28]. The nickel precursors reported are not carbonyl-based; FEBID deposi
based on $Ni(C_5H_4CH_3)_2$ [43,44] or $Ni(PF_3)_4$ [43] do not surpass metal contents of 40 at. %
3D growth of ferromagnets has been attempted with precursors that already provide hig
purity levels in 2D. In the present work, the precursors used are $Co_2(CO)_8$ for cobalt an
$Fe_2(CO)_9$ for iron.

Furthermore, 3D growth implies substantial challenges with respect to 2D growt
As the deposit grows vertically, the geometry of the deposition area changes drasticall
and the relevant electron beam interaction is now with the growing deposit, instead of th
substrate. Many key parameters for the FEBID process are bound to change with respe
to 2D, such as the interaction volume of the electron beam, secondary electron emissio
precursor molecules adsorption and diffusion rates, and heat dissipation [45,46]. Moreove
the surface-to-volume ratio also increases significantly, so surface properties become mo
relevant in 3D nanostructures. This work reviews different possibilities to engineer th
geometrical, compositional and magnetic properties of 3D ferromagnetic nanowires grow
by FEBID, which can be summarized in three main approaches: (1) fine-tuning of FEBI

growth parameters to modify the composition and dimensions [47], (2) the growth of core-shell heterostructures [48], and (3) purification by thermal annealing [49,50].

2. Going 3D: Tuning FEBID Growth Parameters

While 3D FEBID growth has achieved a high degree of architectural complexity with the use of computer-aided design models [17], for the optimization of the physical properties of 3D ferromagnetic deposits, the simple benchmark design of a straight vertical nanowire has been used. For this type of geometry, the FEBID growth is performed in spot mode, where a stationary electron beam is focused into a single point of the substrate for a period of time. To facilitate a detailed, local characterization of the physical properties of the 3D deposits, these are grown out on the edge of a commercial TEM Cu grid.

The main growth parameters left are the primary beam energy, the beam current and the precursor gas flux. While the variation of the primary beam energy by itself does not significantly affect the composition, the interplay between the beam current and the precursor gas flux is essential to determine the final properties of the 3D nanowire. Indeed, two growth modes have been evidenced in Co FEBID nanowires, the so-called linear regime and the radial regime [51]. The transition between these two regimes is marked by a sudden change of the nanowire's diameter, depending on the balance between the beam current and the precursor gas flux. The latter is parameterized through the working pressure, ΔP, defined as the increase of pressure with respect to the base pressure, caused by the precursor gas flux injected during growth. As shown in Figure 1a–d, for a given beam current of 86 pA, a high ΔP of 7.3×10^{-6} mbar produces long nanowires with a diameter well below 75 nm, while at low working pressure ($\Delta P = 5.1 \times 10^{-6}$ mbar), shorter and thicker nanowires are grown, of about 120 nm in diameter. Intermediate values of ΔP give rise to hybrid objects, which evidence linear growth in the early stages up to a certain height, at which the growth transits into radial regime. It is worth noting that for higher beam currents, this transition occurs at higher working pressures, i.e., higher precursor gas fluxes. Thus, the growth rate is determined by the amount of gas molecules delivered (increasing with ΔP), which is known as the precursor-limited regime [52]. Consequently, the nanowires (or segments of nanowire) grown in the linear regime present a high growth rate, expressed in terms of nanowire's length per unit of time, while in the radial regime, nanowires grow more slowly. The growth mode also reflects on the composition of the nanowire. The radial regime gives rise to the nanowires with the highest Co content, close to 90 at. % Co, while in the linear regime, the values decrease below 70 at. % Co. This tendency is observed even for nanowires that present both growth regimes. There is an illustrative example in Figure 1e, which represents the drastic change in Co content at the transition point between linear and radial regime, determined by electron energy loss spectroscopy (EELS) in scanning transmission electron microscopy (STEM).

The microscopic origin of this general behavior is evidently complex and requires careful theoretical simulation. However, some aspects can be qualitatively understood as a consequence of the more or less efficient thermal dissipation, and its impact on the gas precursor molecules adsorption/desorption and decomposition. Considering a nanowire growing in the threshold between the linear and radial regimes, at an early stage of growth, the tip of the nanowire is close to the substrate and the heat generated by the electron beam is easily dissipated onto the substrate. However, as the growth continues, the thermal resistance of the deposit increases, and heat is dissipated less efficiently [45]. At this point, the precursor gas may act as a heat-exchange medium. If the working pressure is high enough, thermal dissipation will be sufficient to maintain the linear growth. However, below a certain working pressure, the temperature at the growth point will increase, favoring a faster decomposition of the gas precursor molecules adsorbed and producing a wider deposit. This subtle balance between the heat produced by the electron probe and the capacity to dissipate it is supported by the fact that, at higher beam current (thus, higher temperature at the growth point), a higher precursor flux (thus, more efficient heat exchange) is required to operate in the linear regime. In terms of metallic content, high gas precursor flux favors an incomplete decomposition of the molecules, decreasin the metallic content of the deposit. As a consequence, high working pressure promotes th growth of narrow, though lower purity, 3D Co nanowires.

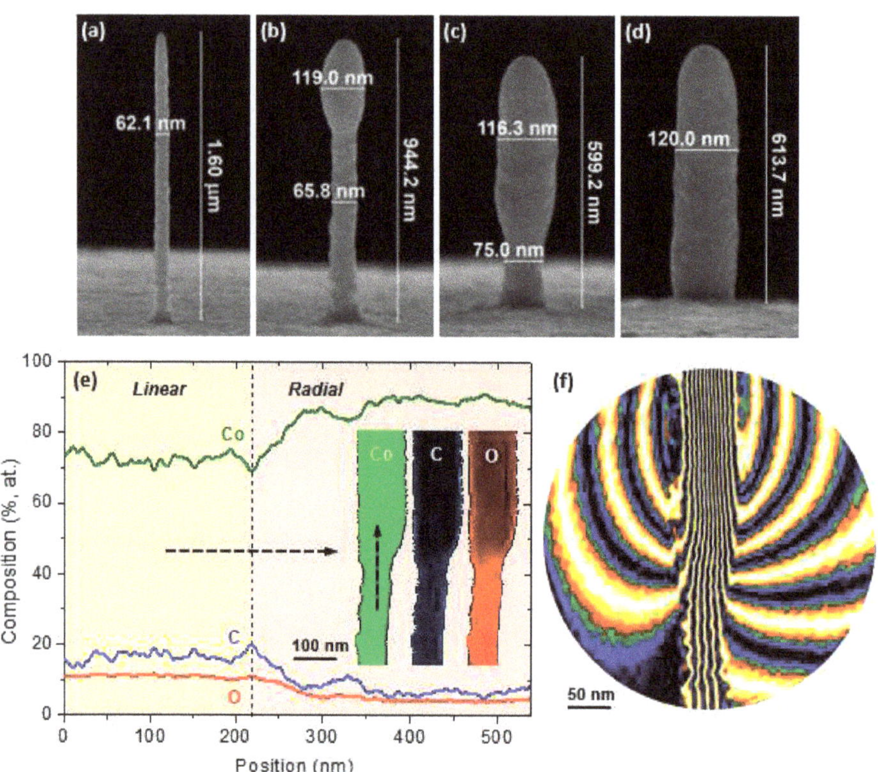

Figure 1. Growth modes of 3D Co focused-electron-beam-induced deposition (FEBID) nanowires. (**a**) Dependence of the transition from linear growth to radial growth with the working pressure increase due to gas injection, ΔP: (**a**) 7.3×10^{-6} mbar, (**b**) 6.4×10^{-6} mbar, (**c**) 5.9×10^{-6} mbar, (**d**) 5.1×10^{-6} mbar. (**e**) Compositional dependence with the growth mode, with STEM-EELS elemental maps in the inset, where Co, C and O are depicted in green, blue and red, respectively. The arrows indicate the direction of the elemental line profile. (**f**) Magnetic flux lines around the transition point between sections grown in radial regime (top) and radial regime (bottom). Adapted from Refs. [47,48].

The beam current is another critical parameter to tailor the composition of 3D nanowire Co content increases with the beam current, sharply at low currents, up to 80 at. % C for 200 pA, and moderately at higher currents. As the beam current rises, the pr cursor molecules are decomposed more efficiently. This also favors the radial regim as the amount of heat to be dissipated increases, and therefore, the diameters tend to b higher [47].

Of course, Co content of the 3D nanowires have a direct impact on the magnetisr This can be analyzed by off-axis electron holography, which is able to quantify the ne magnetic induction of ferromagnetic materials [53]. Figure 1e illustrates qualitatively ho the transition from radial to linear regime in a single nanowire causes a reduction of th magnetic flux lines density, which is associated to the lower magnetic induction (B) cause by the reduced Co content. Figure 2 represents the magnetic induction flux produced b 3D Co nanowires with different diameters. The widest nanowire is grown in the radi; regime (Figure 2a) with a diameter of 124 nm and a composition of 87 at. % Co, and ha an average magnetic induction of 1.33 T, which is 75% of the bulk value (B_{bulk} = 1.76 T). Our estimation does not take into account the fact that the outer surface of the 3D Co FEBID nanowires is oxidized due to air exposure, producing a non-ferromagnetic shell. This oxide layer is highlighted by colored bands in the outer regions where the magnetic induction decays (Figure 2d). The thinnest nanowire is grown in the linear mode (Figure 2c, 57 nm in diameter) and presents a very low magnetic induction of 0.41 T (23% of B_{bulk}), in accordance with a much poorer Co content of 41 at. %. This deposit has a much higher surface-to-volume ratio, so the relative contribution of the oxidized surface is remarkable and it is reasonable to think that the inner magnetic induction values are greatly underestimated. Finally, Figure 2b depicts the nanowire with intermediate thickness, grown in the range where radial and linear regimes coexist. This nanowire evidences values of all physical parameters halfway between the two extreme cases: with a diameter of 81 nm, the composition and magnetic induction values are 68 at. %. Co and 0.78 T (44% of B_{bulk}), respectively.

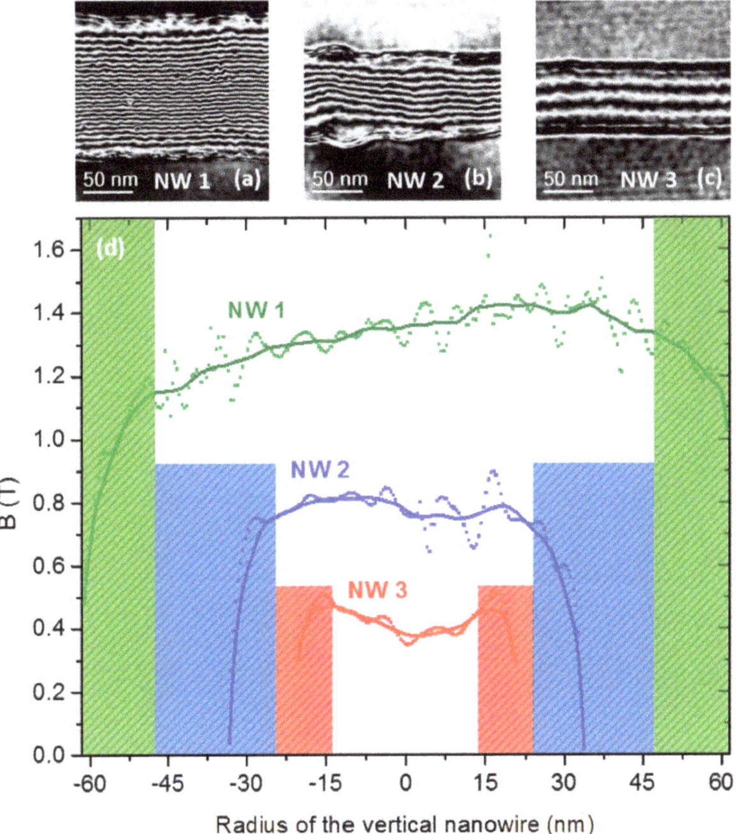

Figure 2. Magnetic properties of as-grown 3D Co FEBID nanowires. (**a–c**) Magnetic induction (B) flux maps of nanowires with diameters of 124, 81 and 57 nm, respectively. (**d**) Cross-sectional profiles of B, where the surface regions are marked with vertical color bands. Adapted from Ref. [47].

This study demonstrates that the variation of the main FEBID growth parameters enables the tailoring of the structural, compositional and magnetic properties of the 3D nanowires. They can even be modulated, as evidenced by the observed change in diameter and composition in a single nanowire at the transition from the linear regime segment

to the radial regime one. This phenomenon could be exploited to engineer pinning
domain walls of exotic nature [54,55]. However, the key growth parameters and physic
properties are mutually dependent, as the Co content, and thus the magnetism, are direct
linked to the growth regime, which determines the average diameter and the growt
rate. These interrelations are summarized in Figure 3, which represents the dependenc
of the Co content of nanowires grown in the optimal conditions for a given diamete
together with the net magnetic induction obtained for the three Co nanowires analyzed t
electron holography. It is clear that both the composition and magnetic induction inevitab
decrease with the reduction of the nanowire's diameter, and the Co content never surpass
70 at. % Co for diameters below 80 nm. According to the holography results, this woul
correspond to a magnetic induction of approximately 0.8 T, which could have a significa
impact on the functionality of the nanowires.

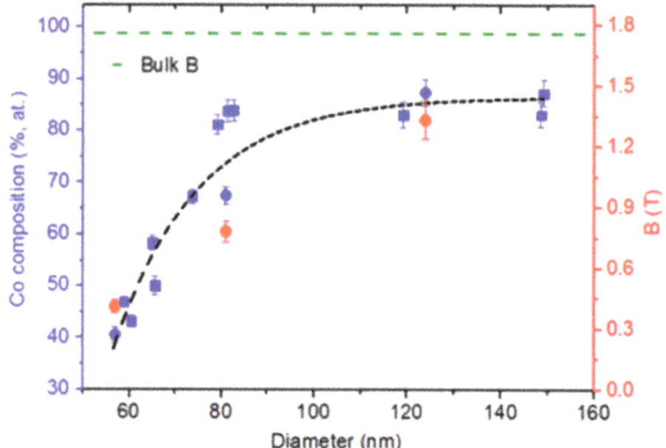

Figure 3. Composition (in blue) and net magnetic induction (in red) of 3D Co nanowires as a functi
of the diameter. Adapted from Ref. [47].

Therefore, to produce 3D Co nanowires with a high aspect ratio, diameters well belo
100 nm and with good functional properties, namely a high (>90 at. %) Co content and
saturation magnetization close to the bulk value, tuning the basic growth parameters is n
enough and post-growth purification procedures are required.

3. Purification by Thermal Annealing

The purification of 2D FEBID deposits with the aim of improving their function
properties has been explored for many years. This matter was exhaustively reviewe
by Botman et al. in 2009 [32] and many works have been reported later. In situ and e
situ thermal annealing in vacuum [39,56,57], or in a reactive atmosphere [58–61], su
strate heating [62,63] or laser irradiation during growth [64,65], post-growth electron bea
irradiation [66,67] or current-induced Joule heating [68], supersonic jet delivery of precu
sor [69], or the exploration of carbon-free precursors [70] are the most relevant methoc
used with different degrees of success. However, in the recent expansion of FEBID
3D [12], few examples of 3D purification can be found in the literature. The most notorio
is the post-growth electron-stimulated purification of 3D gold nanodeposits in a wate
atmosphere to produce virtually pure functional plasmonic nanostructures [21]. In the ca
of ferromagnetic deposits, the most recent efforts have focused on high-vacuum therm
annealing of 3D Co [49] and Fe [50] nanowires, with different results depending on tl
precursor and the as-grown metallic content of the deposits.

As-grown 3D Co nanowires of approximately 90 nm in diameter, an aspect ratio >
and a moderate metallic content of approximately 65 at. % Co were used as starting poi
These nanowires present a pseudo-amorphous nanocrystalline structure with the presen
of a ~5-nm-thick oxide layer (see Figure 4a). Nanowires grown in identical conditions we

fabricated in TEM Cu grids, each one of them annealed at 150, 300, 450 and 600 °C for 100 min in vacuum, with base pressures below 4×10^{-6} mbar. In terms of the morphology, the annealing process preserves the architectural integrity of the nanowires. As can be seen in the last column of Figure 4, there is no perceptible reduction in volume after the annealing process, which contributes to the preservation of the shape and diameter of the nanowires. This is a key ingredient for the application of these procedures in the design of functional devices, and quite remarkable considering the drastic physicochemical transformations that occurred during the annealing process.

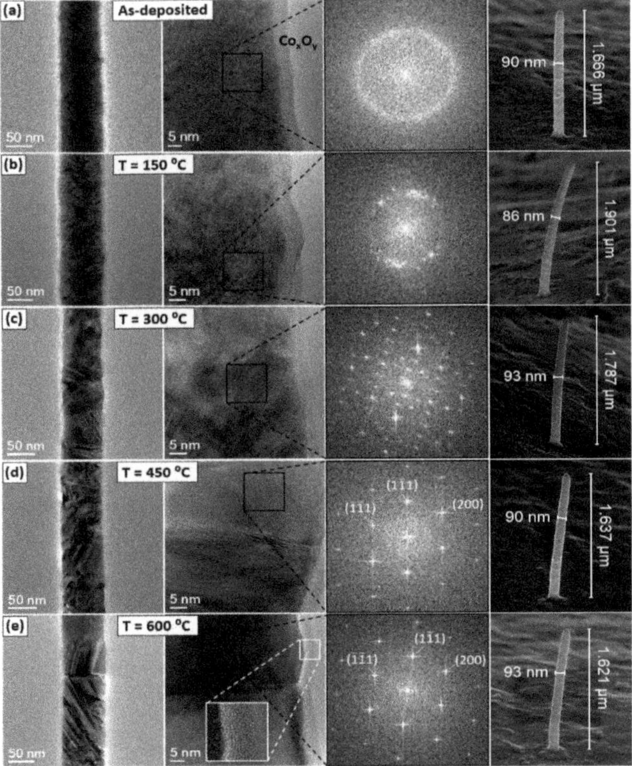

Figure 4. Morphology and microstructure of 3D Co FEBID nanowires as a function of the high-vacuum annealing temperature. TEM, HRTEM images, Fast Fourier Transforms of the squared regions and Scanning Electron Microscopy images of (**a**) an as-grown nanowire and the ones annealed at (**b**) 150 °C, (**c**) 300 °C, (**d**) 450 °C and (**e**) 600 °C. Adapted with permission from Ref. [49]. Copyright 2018 American Chemical Society.

Firstly, the high-vacuum annealing induces the rapid crystallization of the structure. Already at 150 °C, the nanowires evidence the presence of crystals with sizable dimensions, as can be observed in the Fast Fourier Transform of HRTEM images (Figure 4b). Upon further heating, the crystallinity increases (Figure 4c,d) and the average crystal size continues growing, presenting both bcc and hcp structures. Finally, Figure 4e shows how, at 600 °C, only a few Co single crystals that occupy the whole nanowire's diameter remain, while at the surface, a thin layer of partially graphitized carbon remains as a byproduct of the purification.

The elemental composition of the nanowires follows a peculiar trend, which is illustrated in Figure 5. The as-grown homogeneous distribution of Co, C and O becomes

inhomogeneous upon increasing temperature. As the Co content increases, oxygen-ric
(most likely, Co_xO_y) regions nucleate within the nanowire (at temperatures of 300 °C an
450 °C), while carbon accumulates at the surface. At 600 °C, the inner volume is virtual
C- and O-free, with these elements restricted to the surface of the nanowire. This pictu
agrees with the structural model discussed previously of high-purity, highly crystallin
nanowires covered by a thin surface layer of residual contaminants, which gives rise to a
overall metal content of approximately 90 at. % Co.

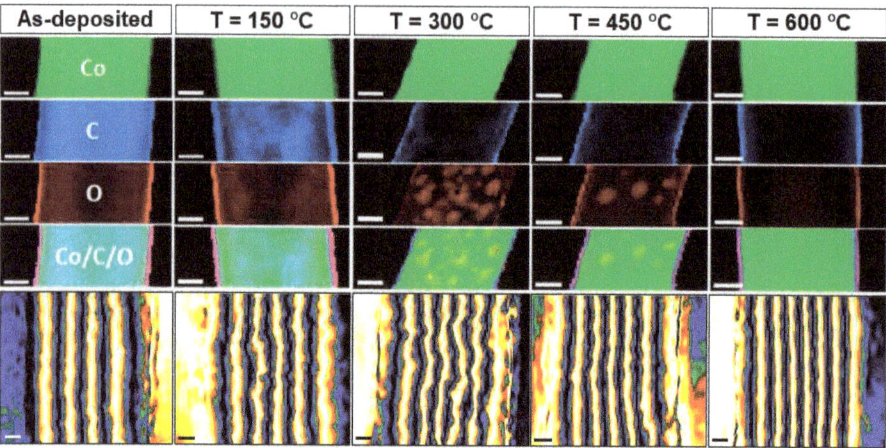

Figure 5. Chemical composition and magnetism of 3D Co FEBID nanowires as a function of the high-vacuum annealing
temperature. The first four rows correspond to STEM-EELS elemental maps of Co, C, and O. The last row plots net magnetic
induction flux maps obtained in the same nanowires. Scale bars are 20 nm in the STEM-EELS images and 10 nm in the
magnetic induction flux maps. Adapted with permission from Ref. [49]. Copyright 2018 American Chemical Society.

Magnetization evolves similarly to the average chemical composition. The ne
magnetic induction of the nanowires increases from as-grown B ~ 0.8 T to B ~ 1.35
for the nanowires annealed at 600 °C, as evidenced from the increasing density of magneti
flux lines in Figure 5. Even though crystallinity has increased, the magnetic configuratio
remains parallel to the nanowire's axis in remanence, so the shape anisotropy is sti
dominant in the purified nanowires.

The elemental analysis of the surficial region changes significantly upon annealin
(not shown here) [49]. While as-grown and low-temperature annealing present an oxyge
rich surface, a form of Co oxide as a consequence of the exposure to air, upon increasin
temperatures, the surface becomes C-rich with a much lower oxygen content. Annealing i
duces the thermal activation of precursor residues, forming volatile species such as C
and CO_2 that migrate to the surface and evaporate [57], leaving behind the graphitized ca
bonous surface, which indeed can be observed in the inset of Figure 4e. This carbonaceou
layer actually serves as a protective layer upon further oxidation in subsequent exposur
to air. In all cases, all the nanowires are covered by a Co-poor, non-magnetic surface whic
depending on the annealing temperature, ranges from 5 to 10 nm. As a consequenc
the composition and magnetic induction values of the inner volume of the nanowires ar
underestimated. After subtracting this contribution and taking into account only the inne
magnetic volume of the nanowires, Figure 6 summarizes how the high-vacuum annealin
process successfully produces virtually pure crystalline and ferromagnetic Co nanowire
with magnetization very close to the bulk value.

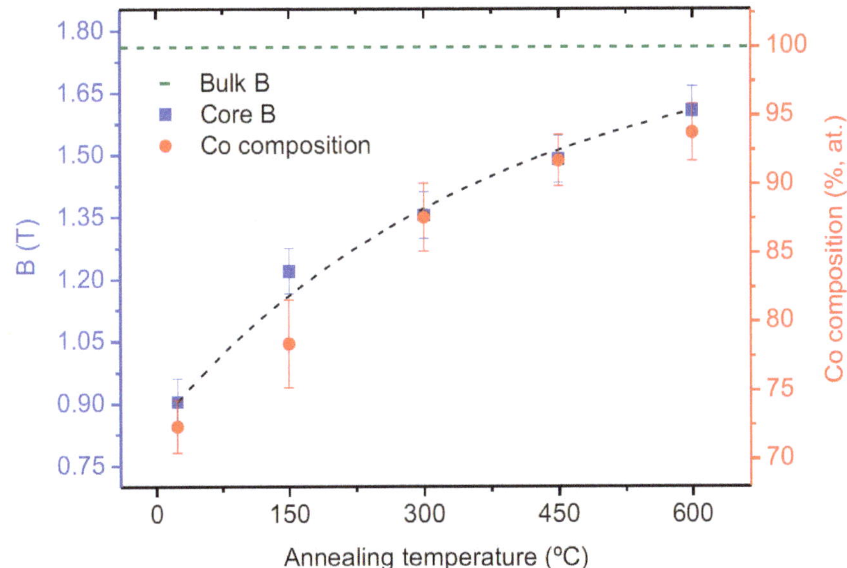

Figure 6. Magnetic induction of the inner magnetic volume and Co composition of 3D Co FEBID nanowires as a function of the annealing temperature. Adapted with permission from Ref. [49]. Copyright 2018 American Chemical Society.

This is not an obvious result. A similar annealing procedure has been conducted in Fe FEBID nanowires, as illustrated in Figure 7. For similar growth conditions, 3D Fe nanowires present narrower dimensions (~50 nm in diameter), higher aspect ratio (>50) and a comparable metallic content of ~75 at. %. In this case, a crystallization process similar to the one in FEBID Co occurs at even lower temperatures (wide crystals are already observed at 450 °C), but the purification is not completely homogeneous. While some areas increase their Fe content, it decreases in other regions, giving rise to an heterogeneous object which, in fact, evidences remarkable changes of shape for an already shorter annealing time of 25 min—shorter than in the Co experiments. This indicates that the structural and chemical changes triggered by thermal annealing are not identical for all the materials grown by FEBID, even if the precursors used are similar. This phenomenon has been further investigated by in situ thermal annealing in a transmission electron microscope, in which the whole purification process upon annealing can be monitored in Fe nanowires with similar dimensions, but a lower metallic content of ~40 at. %. Fe. The in situ characterization of the annealing process reveals that, indeed, an inhomogeneous purification takes place. Figure 8 illustrates how high-purity crystalline Fe regions are formed, interspersed between carbonaceous areas. The low purity of the nanowires in comparison with those used for ex situ annealing aggravates this tendency to heterogeneity, presenting remarkable diameter variations which, combined with the extreme compositional variations, change the architecture dramatically and compromise the structural stability of the nanowire.

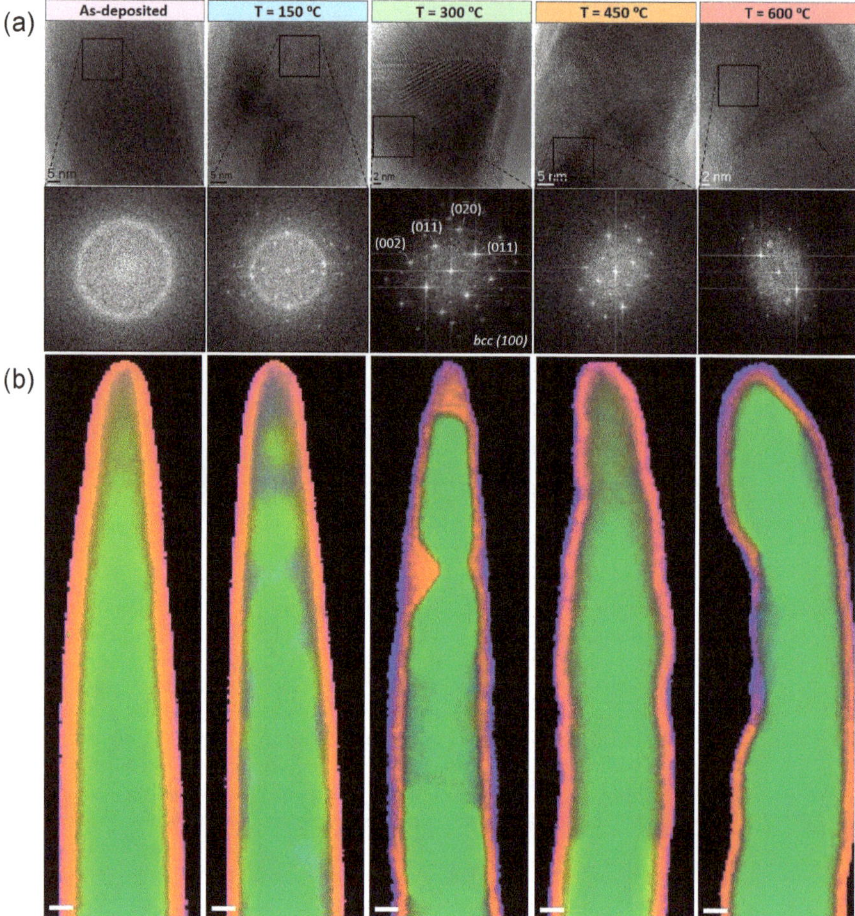

Figure 7. Morphology and microstructure of high-metal-content 3D Fe FEBID nanowires as a function of the high-vacuum annealing temperature. (**a**) HRTEM images and Fast Fourier Transform of the squared regions of the as-grown and annealed objects. (**b**) STEM-EELS elemental maps of the tip of the nanowires, with the spatial distribution of Fe, O and C in green, red and blue, respectively. Undefined scale bars are 10 nm. Reprinted from Ref. [50], Copyright 2019, with permission from Elsevier.

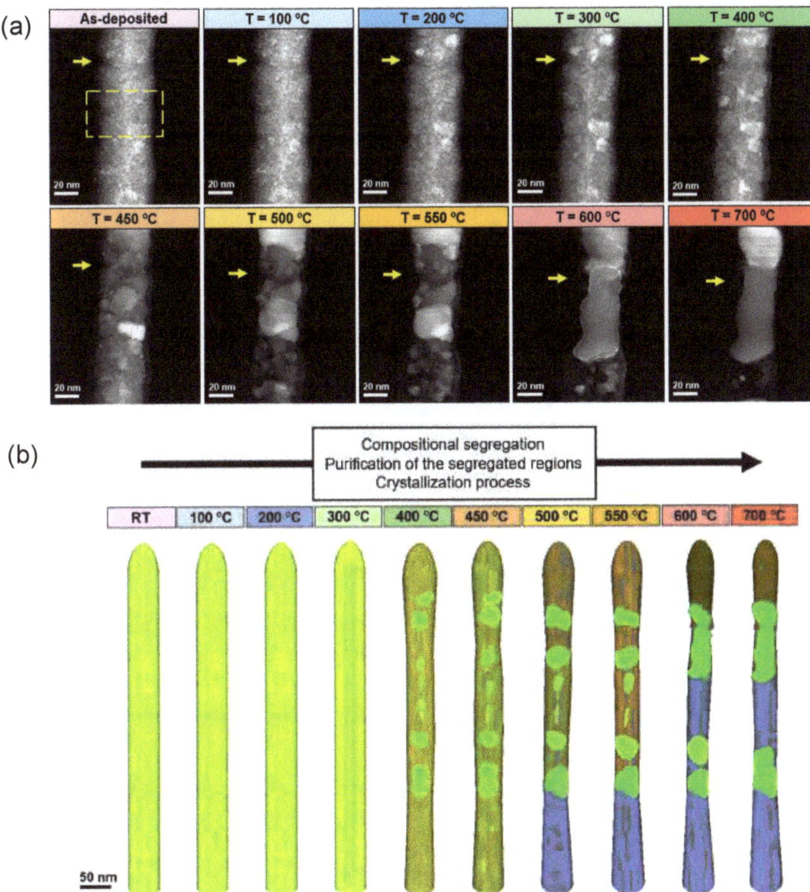

Figure 8. In situ TEM characterization of morphology and microstructure of a low-metal-content 3D Fe FEBID nanowire observed upon high-vacuum thermal annealing. (**a**) STEM images of the same region as a function of temperature. Yellow arrows indicate the same point of the nanowire. (**b**) Sketch of the purification process, where the Fe, O and C spatial distributions are depicted in green, red and blue, respectively. Reprinted from Ref. [50], Copyright 2019, with permission from Elsevier.

The different thermal evolution of 3D Co and Fe raises the question of its microscopic origin. Even though the exact mechanisms underlying the FEBID growth and annealing processes are still unclear, the nature of non-metallic residues, for instance, the C:O ratio, is a good indication of the intervening chemical reactions. Barth et al. concluded that a C:O ratio ≥ 1 might indicate that C-O bonds have been inefficiently cleaved and CO ligands have been incorporated to the deposit, while C:O < 1 suggests the metal oxidation by residual water [71]. In the purification analyses presented in our work, 3D Co corresponds to the first situation, while 3D Fe matches the second one. This is also correlated with the much lower diameter of 3D Fe nanowires; therefore, they are relatively more exposed to ambient atmosphere than Co nanowires. Assuming these considerations, it is very likely that the different chemistry of Co and Fe species in both deposits might be the origin of the different morphological evolution of the nanowires upon thermal annealing.

4. Core-Shell Heterostructures

As the diameter of the 3D nanowires is reduced, the nature of the surface becomes significant in terms of the overall physical properties of the object. As discussed in the previous section, due to the synthetic process and the exposure to ambient atmosphere, a surficial region with a thickness in the range 5–10 nm presents a distinct composition, structure, magnetism and, presumably, transport properties with respect to the core. Both as-grown and annealed nanowires present a surface with much lower metallic content due to natural oxidation or accumulation of residual carbon contaminants upon thermal annealing. This is more critical in FEBID Fe, which tends to produce much thinner nanowires—down to 35 nm in diameter [24]—than in FEBID Co. While in a 100 nm-wide 3D Co nanowire a 5 nm-thick degraded surface represents 19% of the total volume; in a 3D Fe nanowire with a typical diameter of 50 nm, it scales up to 36%. Thus, producing 3D nanowires without surface degradation might be key to obtain functional objects with lateral resolution below 100 nm with optimum, bulk-like physical properties. As annealing does not fully solve this issue, the growth of a protective cover becomes the sole alternative. While different approaches can be envisaged, the most straightforward solution is the direct growth of another layer of material (i.e., a shell) onto the as-grown nanowire (i.e., the core), prior to air exposure. For this purpose, the growth of a Pt-C FEBID shell is the most convenient as $CH_3CpPt(CH_3)_3$ is one of the most standard precursors in SEM-FIB instrumentation and is widely used as a structural or electrical-contact material, with high growth rate and minimal interaction with the deposition area.

Therefore, the growth of core-shell Co@Pt-C and Fe@Pt-C by FEBID has been explored. The growth of a homogeneous shell around the magnetic core introduces an additional step to the synthetic procedure. The only successful procedure has been based on the tilt of the nanowire to lay horizontally with respect to the vertical electron beam, thus allowing the deposition of the shell material using a rectangular pattern corresponding to the length of the nanowire. However, a single-step shell deposition is not sufficient. The delivery of gas precursor is highly directional (depends on the actual position of the gas injector system needle), and a higher abundance of precursor molecules should be expected in the face confronting the injector than in the opposite one [72]. Furthermore, the emission of secondary electrons to decompose the precursor molecules adsorbed will be inherently inhomogeneous along the nanowire's surface, and this compromises the homogeneous coverage of the shell in a single shot [73]. An example of this is shown in Figure 9, which illustrates the attempt to produce a bimetallic core-shell structure by a single-step shell deposition of Co-FEBID onto a 3D Fe FEBID nanowire. The result is that the back side of the nanowire corresponding to the exit surface of the primary beam remains uncovered. Therefore, a second deposition with the nanowire rotated around its symmetry axis by 180° is required to obtain full coverage, as illustrated in Figure 9b for Co@Pt-C FEBID.

Figure 10 qualitatively illustrates the morphology and chemistry of the uncoated 3D Co and Fe cores with respect to their associated Pt-C shell covered counterparts. From the microstructure point of view, the naked magnetic cores, see Figure 10a,c, present the typical nanocrystalline structure covered by a low-density, metal-poor surface. The elemental maps evidence that this is the surficial oxide layer due to air exposure, signaled by the thin metal oxide layer depicted in orange color. In the case of the core-shell structures depicted in Figure 10b,d, the nature of the core is masked by the typical microstructure of FEBID Pt-C, which is an amorphous carbonaceous matrix with small Pt nanoparticles embedded. The distribution of chemical elements of the core-shell structure is radically different from the uncoated ones, as shown by the full green (high metallic content) inner part of the nanowires covered by a vivid red shell, free of Co or Fe. The lack of a transitional region (orangish) between core and shell evidence that there is no Co surface degradation during or after the synthetic process. The oxygen content is also remarkably uniform across the core [48], as these oxygen contaminants are derived from incomplete precursor molecule decomposition and not a result of surface degradation. The quasi-cylindrical geometry of the Co core and the homogeneous coverage of the Pt-C coverage achieved by the two-step shell growth, which was already hinted by the cross-sectional cut shown in Figure 9b, is confirmed all along the nanowire.

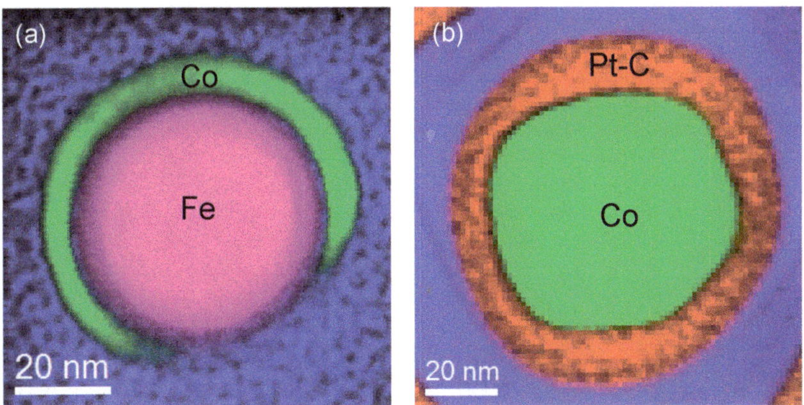

Figure 9. Optimization of 3D core-shell FEBID nanowires. (**a**) Cross-section of a one-step deposition of FEBID Fe@Co. (**b**) Cross-section of a two-step deposition of Co@Pt-C. Adapted from Refs. [48,74].

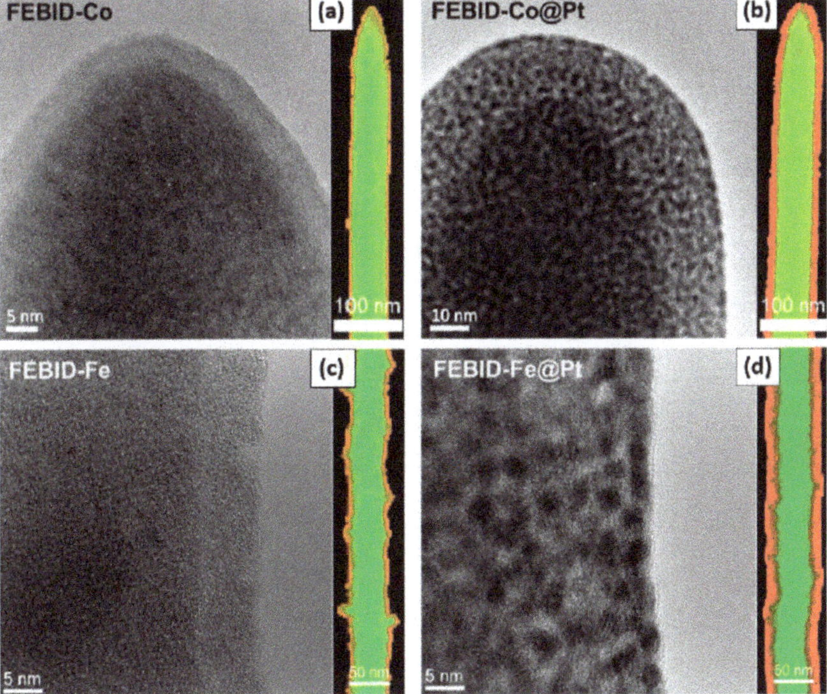

Figure 10. Microstructure and chemical composition of 3D core-shell (**b**) Co@Pt-C and (**d**) Fe@Pt-C FEBID nanowires with respect to the uncoated (**a**) Co and (**c**) Fe nanowires. Each image is accompanied by a STEM-EELS elemental map of the central part of the nanowires where Co(Fe), and O are depicted in green and red, respectively. Adapted from Ref. [48].

The most prominent impact of the core-shell architecture is in the magnetic properties. Again, electron holography is used to perform a quantitative analysis of the net magnetic

induction produced by each 3D nanowire [36]. Figure 11 depicts the net magnetic inductic
flux produced by the set of four 3D nanowire structures and their spatial distributic
This magnetic flux has been quantified, assuming a perfect cylindrical symmetry ar
normalizing by the nominal diameter of the core. Firstly, holographic images of tl
uncoated nanowires indicate the suppression or weakening of ferromagnetism in tl
uncoated nanowires due to air exposure and oxidation—see Figure 11a,d. This is evidenc
by the disappearance (in Fe) or decreased density (in Co) of magnetic flux at the surfa
Secondly, the magnetic flux of the cores increases with respect to the uncoated nanowires
see Figure 11b,e. This is clearly displayed in the magnetic induction profiles integrate
across the nanowires' diameter, as shown in Figure 11c,f. The increase in magnetic inductic
is notable, about 20% in the case of Co and 35% in Fe. This values still remain far fro
bulk values—1.0 T vs. 1.76 T in Co; 1.25 T vs. 2.2 T in Fe. This discrepancy is, howeve
due to the minute diameter of the cores of the nanowires selected for the magnetic stuc
which gives rise to particularly low metallic contents in the core, as discussed in Section :

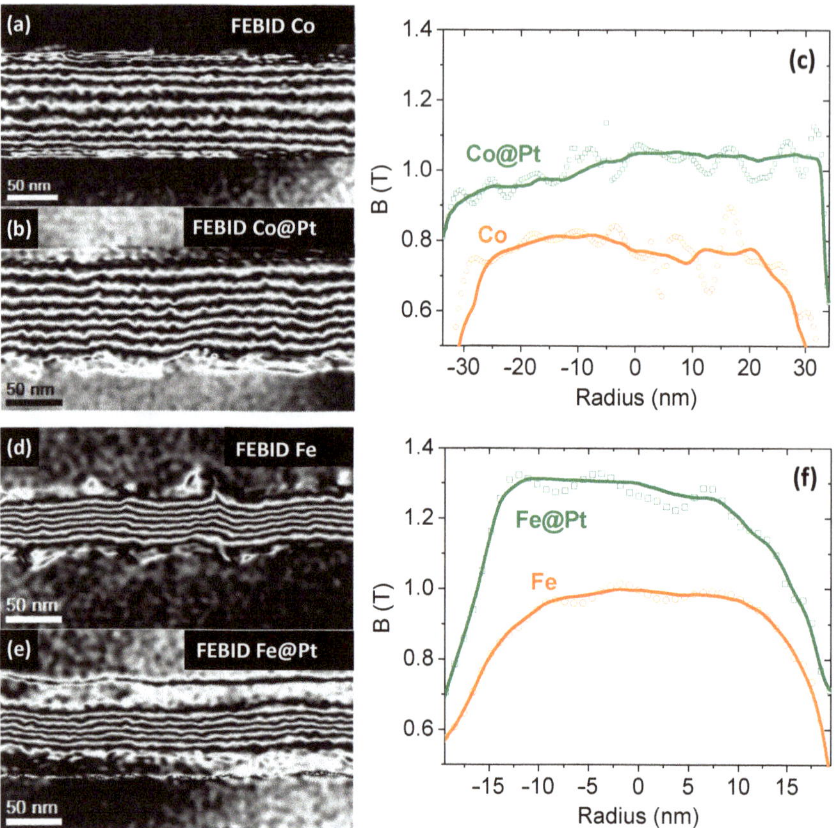

Figure 11. Magnetism of 3D core-shell (**b**) Co@Pt-C and (**e**) Fe@Pt-C FEBID nanowires with respect to the uncoated (**a**) Co and (**d**) Fe nanowires. Magnetic flux maps of each nanowire in the left panel are illustrated by cross-sectional profiles of the net magnetic induction of corresponding (**c**) Co and (**f**) Fe nanowires in the right panel. Adapted from Ref. [48].

5. Conclusions and Perspectives

FEBID is an additive nano-manufacturing technique which offers great versatility
fabricate complex motifs and architectural designs at the nanoscale based on numerous n
terials, and in particular, ferromagnetic materials for high-density, low-power applicatio

as memories, sensors and actuators. The possibilities of 2D FEBID ferromagnetic deposits have been extensively explored, but the expansion to 3D ferromagnetic nanostructures has encountered new challenges in terms of lateral resolution, metallic content and, therefore, the quality of functional properties. Some key advances in the control of growth and optimization of the functional properties of 3D ferromagnetic nanostructures synthesized by FEBID have been reviewed here, using vertical straight nanowires of high aspect ratio as the benchmark.

Understanding the FEBID growth processes of 3D geometries is key to customize their properties. In this regard, one of the main results obtained is the existence of two growth regimes that depend on the subtle balance between the electron beam current and the precursor gas flux, which affects the capacity of the growing nanostructure to dissipate heat during FEBID growth. Depending on these parameters, a radial regime is predominant for low precursor gas flux in which high purity is obtained (>80 at. % Co), but the growth rate is low and the lateral dimensions cannot go below 100 nm in diameter. Higher precursor fluxes enable faster growth, in terms of nanowire length for a given deposition time, and diameters well below 100 nm; however, the metallic content decreases down to 40–50 at. % Co for diameters around 50 nm and magnetization diminishes accordingly. Thus, purification procedures might be required to obtain functional 3D nanostructures with lateral resolutions below 100 nm. High-vacuum thermal annealing has been reviewed as a straightforward approach, as it could be easily implemented right on the deposition chamber. Dissimilar results have been obtained depending on the precursor gas employed and the starting metallic content. 3D Co nanowires based on $Co_2(CO)_8$ precursor achieve virtually pure, crystalline and homogeneous nanowires with bulk-like ferromagnetic properties, with minimum volume reduction, which preserves the original architecture. On the other hand, 3D Fe nanowires grown with $Fe_2(CO)_9$ show a tendency toward phase segregation upon annealing, displaying a mixture of high-purity metal segments interspersed with carbonaceous areas. Both kinds present very different mass density and mechanical strengths, which cause serious shrinkage, drastic diameter variations and, therefore, architectural instability.

The role of surface has been also extensively analysed, and a remarkable degradation of the surface has been observed, mostly due to ambient air exposure. In high surface-to-volume ratio nanostructures such as 3D nanowires, the overall properties depend on surface properties and quality, so a procedure to deposit a Pt-C protective cover by FEBID right after Co or Fe growth has been described. As a result, core-shell nanowires have been designed in which the ferromagnetic cores are free of surface deterioration, with the subsequent improvement in the functional properties, especially in ultrathin nanowires (down to 35 nm in diameter) in which up to 1/2 of the nanowire's volume is affected by air oxidation.

These findings open new pathways to produce high-quality 3D ferromagnetic nanowires with bulk-like properties and lateral resolution well below 100 nm, and to continue optimizing not only the growth of 3D architectures of nanomagnets, but also of non-magnetic materials. The combination of smart tuning of the growth conditions and thermal annealing in 3D FEBID Co enables the possibility of focusing on the architectural design with the smallest possible lateral dimensions, faster growth and high aspect ratios, at the expense of an optimal composition, which can be improved afterwards by thermal annealing or other methods yet to be explored. The existence of two growth regimes also stimulates the idea of in situ variation growth conditions to produce 3D nanostructures with modulated diameters [75,76]. This idea could be complemented by other strategies that have been recently proposed for diameter reduction and modulation of 3D nanowires, such as the application of local electric fields or the live control of electron beam focus [77]. In cases where post-growth thermal annealing causes severe shrinkage, other means of purification should be explored further to grant architectural stability. For instance, electron stimulation upon reactive atmospheres has been successful in FEBID Au nanostructures [21], and could be explored in Co and Fe, during or after growth.

Once the possibility of fabricating 3D heterostructures by FEBID has been demonstrated, a vast field of research will emerge for optimizing the procedure and inventing novel complex architectures based on the combination of two or more layers or segments of material [54]. In the field of nanomagnetism, the strategy followed for Co@Pt-nanowires can be reversed, using the same materials to produce ferromagnetic nanotubes on non-magnetic templates, for which attractive magnetic properties for spintronic devices have been predicted in terms of domain wall conduit speed and stability [9] due to the lack of a magnetic core. We have obtained preliminary results indicating that Co FEBID nanotubes are ferromagnetic and nucleate exotic domain walls [74]. The growth of bimetallic nanowires combining Co, Fe and Ni precursors has been hinted in this work and needs to be explored as a mean to introduce an intrinsic modulation of magnetic properties into the 3D nanomagnet. High-quality interfaces between ferromagnetic materials and high spin-orbit coupling materials such as Pt could be studied for charge-spin conversion in spin-orbitronic applications [78], if proper purification procedures can be developed for 3D FEBID Pt [59]. Considering the available materials, the combination of insulating, conducting, ferromagnetic, superconducting or plasmonic materials is possible to investigate exciting phenomena such as proximity effects with superconductors, designing magneto-optically active nano-objects by combination with plasmonic metals, etc. Finally, the combination of FEBID with other synthetic techniques such as atomic layer deposition could provide infinite possibilities for nano-architectural design of 3D ferromagnetic nanostructures.

Author Contributions: The manuscript was conceived and written by C.M. C.M., J.P.-N. and J.M.D. contributed to the experimental results, data analysis and interpretation. C.M., J.P.-N. and J.M.D. have reviewed and corrected the manuscript. All authors have read and agreed to the published version of the manuscript.

Funding: This research was funded by the Spanish Ministry of Economy and Competitiveness through the projects MAT2017-82970-C2-1-R and MAT2017-82970-C2-2-R, the Aragon Regional Government (Construyendo Europa desde Aragón) through the project E13_20R with European Social Fund funding, the Ayuda para Contratos Predoctorales para la Formación de Doctores (BES 2015-072950) of the Spanish MINECO with the participation of the European Social Fund, and the European's Union Horizon 2020 research and innovation programme under Grant No. 82371 ESTEEM3.

Conflicts of Interest: The authors declare no conflict of interest.

References

1. Fernández-Pacheco, A.; Streubel, R.; Fruchart, O.; Hertel, R.; Fischer, P.; Cowburn, R.P. Three-dimensional nanomagnetism *Nat. Commun.* **2017**, *8*, 15756. [CrossRef]
2. Streubel, R.; Fischer, P.; Kronast, F.; Kravchuk, V.P.; Sheka, D.D.; Gaididei, Y.; Schmidt, O.G.; Makarov, D. Magnetism in curved geometries. *J. Phys. D Appl. Phys.* **2016**, *49*, 363001. [CrossRef]
3. Staňo, M.; Fruchart, O. Magnetic Nanowires and Nanotubes. In *Handbook of Magnetic Materials*; Brück, E., Ed.; Elsevier: Amsterdam, The Netherlands, 2018; Volume 27, pp. 155–267. ISBN 9780444641618.
4. Berganza, E.; Jaafar, M.; Fernandez-Roldan, J.A.; Goiriena-Goikoetxea, M.; Pablo-Navarro, J.; García-Arribas, A.; Guslienko, K.; Magén, C.; De Teresa, J.M.; Chubykalo-Fesenko, O.; et al. Half-hedgehog spin textures in sub-100 nm soft magnetic nanodots. *Nanoscale* **2020**, *12*, 18646–18653. [CrossRef]
5. Parkin, S.S.P.; Hayashi, M.; Thomas, L. Magnetic Domain-Wall Racetrack Memory. *Science* **2008**, *320*, 190–194. [CrossRef]
6. Brataas, A.; Kent, A.D.; Ohno, H. Current-induced torques in magnetic materials. *Nat. Mater.* **2012**, *11*, 372–381. [CrossRef] [PubMed]
7. Puebla, J.; Kim, J.; Kondou, K.; Otani, Y. Spintronic devices for energy-efficient data storage and energy harvesting. *Commun. Mater.* **2020**, *1*, 24. [CrossRef]
8. Radamson, H.H.; Zhu, H.; Wu, Z.; He, X.; Lin, H.; Liu, J.; Xiang, J.; Kong, Z.; Xiong, W.; Li, J.; et al. State of the Art and Future Perspectives in Advanced CMOS Technology. *Nanomaterials* **2020**, *10*, 1555. [CrossRef] [PubMed]
9. Hertel, R. Ultrafast domain wall dynamics in magnetic nanotubes and nanowires. *J. Phys. Condens. Matter* **2016**, *28*, 48300. [CrossRef] [PubMed]
10. Yan, M.; Kákay, A.; Gliga, S.; Hertel, R. Beating the Walker limit with massless domain walls in cylindrical nanowires. *Phys. Rev. Lett.* **2010**, *104*, 057201. [CrossRef] [PubMed]

Fert, A.; Reyren, N.; Cros, V. Magnetic skyrmions: Advances in physics and potential applications. *Nat. Rev. Mater.* **2017**, *2*, 17031. [CrossRef]

Winkler, R.; Fowlkes, J.D.; Rack, P.D.; Plank, H. 3D nanoprinting via focused electron beams. *J. Appl. Phys.* **2019**, *125*, 210901. [CrossRef]

Utke, I.; Hoffmann, P.; Melngailis, J. Gas-assisted focused electron beam and ion beam processing and fabrication. *J. Vac. Sci. Technol. B Microelectron. Nanom. Struct.* **2008**, *26*, 1197–1276. [CrossRef]

Van Dorp, W.F.; Hagen, C.W. A critical literature review of focused electron beam induced deposition. *J. Appl. Phys.* **2008**, *104*, 081301. [CrossRef]

Fowlkes, J.D.; Randolph, S.J.; Rack, P.D. Growth and simulation of high-aspect ratio nanopillars by primary and secondary electron-induced deposition. *J. Vac. Sci. Technol. B Microelectron. Nanom. Struct.* **2005**, *23*, 2825. [CrossRef]

Ven Kouwen, L.; Botman, A.; Hagen, C.W. Focused electron-Beam-induced deposition of 3 nm dots in a scanning electron microscope. *Nano Lett.* **2009**, *9*, 2149–2152. [CrossRef] [PubMed]

Fowlkes, J.D.; Winkler, R.; Lewis, B.B.; Stanford, M.G.; Plank, H.; Rack, P.D. Simulation-Guided 3D Nanomanufacturing via Focused Electron Beam Induced Deposition. *ACS Nano* **2016**, *10*, 6163–6172. [CrossRef] [PubMed]

Sanz-Hernández, D.; Hamans, R.; Osterrieth, J.; Liao, J.-W.; Skoric, L.; Fowlkes, J.; Rack, P.; Lippert, A.; Lee, S.; Lavrijsen, R.; et al. Fabrication of Scaffold-Based 3D Magnetic Nanowires for Domain Wall Applications. *Nanomaterials* **2018**, *8*, 483. [CrossRef] [PubMed]

Brintlinger, T.; Fuhrer, M.S.; Melngailis, J.; Utke, I.; Bret, T.; Perentes, A.; Hoffmann, P.; Abourida, M.; Doppelt, P. Electrodes for carbon nanotube devices by focused electron beam induced deposition of gold. *J. Vac. Sci. Technol. B Microelectron. Nanom. Struct.* **2005**, *23*, 3174. [CrossRef]

Plank, H.; Winkler, R.; Schwalb, C.H.; Hütner, J.; Fowlkes, J.D.; Rack, P.D.; Utke, I.; Huth, M. Focused Electron Beam-Based 3D Nanoprinting for Scanning Probe Microscopy: A Review. *Micromachines* **2019**, *11*, 48. [CrossRef]

Winkler, R.; Schmidt, F.P.; Haselmann, U.; Fowlkes, J.D.; Lewis, B.B.; Kothleitner, G.; Rack, P.D.; Plank, H. Direct-Write 3D Nanoprinting of Plasmonic Structures. *ACS Appl. Mater. Interfaces* **2017**, *9*, 8233–8240. [CrossRef]

Stiller, M.; Barzola-Quiquia, J.; Esquinazi, P.D.; Sangiao, S.; De Teresa, J.M.; Meijer, J.; Abel, B. Functionalized Akiyama tips for magnetic force microscopy measurements. *Meas. Sci. Technol.* **2017**, *28*, 125401. [CrossRef]

Sangiao, S.; Magén, C.; Mofakhami, D.; de Loubens, G.; De Teresa, J.M. Magnetic properties of optimized cobalt nanospheres grown by focused electron beam induced deposition (FEBID) on cantilever tips. *Beilstein J. Nanotechnol.* **2017**, *8*, 2106–2115. [CrossRef] [PubMed]

Jaafar, M.; Pablo-Navarro, J.; Berganza, E.; Ares, P.; Magén, C.; Masseboeuf, A.; Gatel, C.; Snoeck, E.; Gómez-Herrero, J.; de Teresa, J.M.; et al. Customized MFM probes based on magnetic nanorods. *Nanoscale* **2020**, *12*, 10090–10097. [CrossRef] [PubMed]

Vavassori, P.; Pancaldi, M.; Perez-Roldan, M.J.; Chuvilin, A.; Berger, A. Remote Magnetomechanical Nanoactuation. *Small* **2016**, *12*, 1013–1023. [CrossRef] [PubMed]

Sanz-Hernández, D.; Hamans, R.F.; Liao, J.W.; Welbourne, A.; Lavrijsen, R.; Fernández-Pacheco, A. Fabrication, Detection, and Operation of a Three-Dimensional Nanomagnetic Conduit. *ACS Nano* **2017**, *11*, 11066–11073. [CrossRef] [PubMed]

Gavagnin, M.; Wanzenboeck, H.D.; Wachter, S.; Shawrav, M.M.; Persson, A.; Gunnarsson, K.; Svedlindh, P.; Stöger-Pollach, M.; Bertagnolli, E. Free-standing magnetic nanopillars for 3D nanomagnet logic. *ACS Appl. Mater. Interfaces* **2014**, *6*, 20254–20260. [CrossRef] [PubMed]

Keller, L.; Al Mamoori, M.K.I.; Pieper, J.; Gspan, C.; Stockem, I.; Schröder, C.; Barth, S.; Winkler, R.; Plank, H.; Pohlit, M.; et al. Direct-write of free-form building blocks for artificial magnetic 3D lattices. *Sci. Rep.* **2018**, *8*, 6160. [CrossRef]

Al Mamoori, M.K.I.; Keller, L.; Pieper, J.; Barth, S.; Winkler, R.; Plank, H.; Müller, J.; Huth, M. Magnetic characterization of direct-write free-form building blocks for artificial magnetic 3D lattices. *Materials* **2018**, *11*, 289. [CrossRef]

Sanz-Hernández, D.; Hierro-Rodriguez, A.; Donnelly, C.; Pablo-Navarro, J.; Sorrentino, A.; Pereiro, E.; Magén, C.; McVitie, S.; De Teresa, J.M.; Ferrer, S.; et al. Artificial Double-Helix for Geometrical Control of Magnetic Chirality. *ACS Nano* **2020**. [CrossRef]

Fernández-Pacheco, A.; Skoric, L.; De Teresa, J.M.; Pablo-Navarro, J.; Huth, M.; Dobrovolskiy, O.V. Writing 3D Nanomagnets Using Focused Electron Beams. *Materials* **2020**, *13*, 3774. [CrossRef]

Botman, A.; Mulders, J.J.L.; Hagen, C.W. Creating pure nanostructures from electron-beam-induced deposition using purification techniques: A technology perspective. *Nanotechnology* **2009**, *20*, 372001. [CrossRef]

Arnold, G.; Timilsina, R.; Fowlkes, J.; Orthacker, A.; Kothleitner, G.; Rack, P.D.; Plank, H. Fundamental Resolution Limits during Electron-Induced Direct-Write Synthesis. *ACS Appl. Mater. Interfaces* **2014**, *6*, 7380–7387. [CrossRef] [PubMed]

De Teresa, J.M.; Fernández-Pacheco, A.; Córdoba, R.; Serrano-Ramón, L.; Sangiao, S.; Ibarra, M.R. Review of magnetic nanostructures grown by focused electron beam induced deposition (FEBID). *J. Phys. D Appl. Phys.* **2016**, *49*, 243003. [CrossRef]

Fernández-Pacheco, A.; De Teresa, J.M.; Córdoba, R.; Ibarra, M.R.; Fernndez-Pacheco, A.; De Teresa, J.M.; Córdoba, R.; Ibarra, M.R. Magnetotransport properties of high-quality cobalt nanowires grown by focused-electron-beam-induced deposition. *J. Phys. D Appl. Phys.* **2009**, *42*, 055005. [CrossRef]

Serrano-Ramón, L.; Córdoba, R.; Rodríguez, L.A.; Magén, C.; Snoeck, E.; Gatel, C.; Serrano, I.; Ibarra, M.R.; De Teresa, J.M. Ultrasmall functional ferromagnetic nanostructures grown by focused electron-beam-induced deposition. *ACS Nano* **2011**, *5*, 7781–7787. [CrossRef]

Gazzadi, G.C.; Mulders, H.; Trompenaars, P.; Ghirri, A.; Affronte, M.; Grillo, V.; Frabboni, S. Focused Electron Beam Deposition of Nanowires from Cobalt Tricarbonyl Nitrosyl (Co(CO) 3 NO) Precursor. *J. Phys. Chem. C* **2011**, *115*, 19606–19611. [CrossRef]

38. Rosenberg, S.G.; Barclay, M.; Fairbrother, D.H. Electron Beam Induced Reactions of Adsorbed Cobalt Tricarbonyl Nitrosyl (Co(CO)3 NO) Molecules. *J. Phys. Chem. C* **2013**, *117*, 16053–16064. [CrossRef]
39. Shimojo, M.; Takeguchi, M.; Tanaka, M.; Mitsuishi, K.; Furuya, K. Electron beam-induced deposition using iron carbonyl and the effects of heat treatment on nanostructure. *Appl. Phys. A Mater. Sci. Process.* **2004**, *79*, 1869–1872. [CrossRef]
40. Lavrijsen, R.; Córdoba, R.; Schoenaker, F.J.; Ellis, T.H.; Barcones, B.; Kohlhepp, J.T.; Swagten, H.J.M.M.; Koopmans, De Teresa, J.M.; Magén, C.; et al. Fe:O:C grown by focused-electron-beam-induced deposition: Magnetic and electric properti *Nanotechnology* **2011**, *22*, 025302. [CrossRef]
41. Lukasczyk, T.; Schirmer, M.; Steinrück, H.P.; Marbach, H. Electron-beam-induced deposition in ultrahigh vacuum: Lith graphic fabrication of clean iron nanostructures. *Small* **2008**, *4*, 841–846. [CrossRef]
42. Porrati, F.; Pohlit, M.; Müller, J.; Barth, S.; Biegger, F.; Gspan, C.; Plank, H.; Huth, M. Direct writing of CoFe alloy nanostructu by focused electron beam induced deposition from a heteronuclear precursor. *Nanotechnology* **2015**, *26*, 475701. [CrossRef]
43. Perentes, A.; Sinicco, G.; Boero, G.; Dwir, B.; Hoffmann, P. Focused electron beam induced deposition of nickel. *J. Vac. Sci. Techn B Microelectron. Nanom. Struct.* **2007**, *25*, 2228. [CrossRef]
44. Córdoba, R.; Barcones, B.; Roelfsema, E.; Verheijen, M.A.; Mulders, J.J.L.; Trompenaars, P.H.F.; Koopmans, B. Functional nick based deposits synthesized by focused beam induced processing. *Nanotechnology* **2016**, *27*, 065303. [CrossRef]
45. Mutunga, E.; Winkler, R.; Sattelkow, J.; Rack, P.D.; Plank, H.; Fowlkes, J.D. Impact of Electron-Beam Heating during 3 Nanoprinting. *ACS Nano* **2019**, *13*, 5198–5213. [CrossRef] [PubMed]
46. Skoric, L.; Sanz-Hernández, D.; Meng, F.; Donnelly, C.; Merino-Aceituno, S.; Fernández-Pacheco, A. Layer-by-Lay Growth of Complex-Shaped Three-Dimensional Nanostructures with Focused Electron Beams. *Nano Lett.* **2020**, *20*, 184–1 [CrossRef] [PubMed]
47. Pablo-Navarro, J.; Sanz-Hernández, D.; Magén, C.; Fernández-Pacheco, A.; Teresa, J.M. De Tuning shape, composition a magnetization of 3D cobalt nanowires grown by focused electron beam induced deposition (FEBID). *J. Phys. D Appl. Phys.* **20** *50*, 18LT01. [CrossRef]
48. Pablo-Navarro, J.; Magén, C.; de Teresa, J.M. Three-dimensional core–shell ferromagnetic nanowires grown by focused electro beam induced deposition. *Nanotechnology* **2016**, *27*, 285302. [CrossRef]
49. Pablo-Navarro, J.; Magén, C.; De Teresa, J.M. Purified and Crystalline Three-Dimensional Electron-Beam-Induced Deposi The Successful Case of Cobalt for High-Performance Magnetic Nanowires. *ACS Appl. Nano Mater.* **2018**, *1*, 38–46. [CrossRef]
50. Pablo-Navarro, J.; Winkler, R.; Haberfehlner, G.; Magén, C.; Plank, H.; De Teresa, J.M. *In situ* real-time annealing of ultrath vertical Fe nanowires grown by focused electron beam induced deposition. *Acta Mater.* **2019**, *174*, 379–386. [CrossRef]
51. Hochleitner, G.; Wanzenboeck, H.D.; Bertagnolli, E. Electron beam induced deposition of iron nanostructures. *J. Vac. Sci. Techn B Microelectron. Nanom. Struct.* **2008**, *26*, 939. [CrossRef]
52. Wachter, S.; Gavagnin, M.; Wanzenboeck, H.D.; Shawrav, M.M.; Belić, D.; Bertagnolli, E. Nitrogen as a carrier gas for regin control in focused electron beam induced deposition. *Nanofabrication* **2014**, *1*, 16–22. [CrossRef]
53. Gatel, C.; Snoeck, E. Magnetic mapping using electron holography. In *Transmission Electron Microscopy in Micro-Nanoelectron* Claverie, A., Ed.; ISTE-Wiley: London, UK, 2012.
54. Ivanov, Y.P.; Chuvilin, A.; Lopatin, S.; Kosel, J. Modulated Magnetic Nanowires for Controlling Domain Wall Motion: Toward 3 Magnetic Memories. *ACS Nano* **2016**, *10*, 5326–5332. [CrossRef]
55. Fernandez-Roldan, J.A.; Perez del Real, R.; Bran, C.; Vazquez, M.; Chubykalo-Fesenko, O. Magnetization pinning in modulat nanowires: From topological protection to the "corkscrew" mechanism. *Nanoscale* **2018**, *10*, 5923–5927. [CrossRef] [PubMed]
56. Tanaka, M.; Shimojo, M.; Takeguchi, M.; Mitsuishi, K.; Furuya, K. Formation of iron nano-dot arrays by electron beam-induc deposition using an ultrahigh vacuum transmission electron microscope. *J. Cryst. Growth* **2005**, *275*, 2361–2366. [CrossRef]
57. Puydinger dos Santos, M.V.; Velo, M.F.; Domingos, R.D.; Zhang, Y.; Maeder, X.; Guerra-Nuñez, C.; Best, J.P.; Béron, F.; Pirota, K. Moshkalev, S.; et al. Annealing-Based Electrical Tuning of Cobalt–Carbon Deposits Grown by Focused-Electron-Beam-Induc Deposition. *ACS Appl. Mater. Interfaces* **2016**, *8*, 32496–32503. [CrossRef] [PubMed]
58. Takeguchi, M.; Shimojo, M.; Furuya, K. Nanostructure Fabrication by Electron-Beam-Induced Deposition with Metal Carbor Precursor and Water Vapor. *Jpn. J. Appl. Phys.* **2007**, *46*, 6183–6186. [CrossRef]
59. Plank, H.; Noh, J.H.; Fowlkes, J.D.; Lester, K.; Lewis, B.B.; Rack, P.D. Electron-Beam-Assisted Oxygen Purification at Lo Temperatures for Electron-Beam-Induced Pt Deposits: Towards Pure and High-Fidelity Nanostructures. *ACS Appl. Mat Interfaces* **2014**, *6*, 1018–1024. [CrossRef]
60. Belić, D.; Shawrav, M.M.; Gavagnin, M.; Stöger-Pollach, M.; Wanzenboeck, H.D.; Bertagnolli, E. Direct-Write Depositi and Focused-Electron-Beam-Induced Purification of Gold Nanostructures. *ACS Appl. Mater. Interfaces* **2015**, *7*, 2467–24 [CrossRef] [PubMed]
61. Begun, E.; Dobrovolskiy, O.V.; Kompaniiets, M.; Sachser, R.; Gspan, C.; Plank, H.; Huth, M. Post-growth purification of C nanostructures prepared by focused electron beam induced deposition. *Nanotechnology* **2015**, *26*, 075301. [CrossRef] [PubMed]
62. Córdoba, R.; Sesé, J.; De Teresa, J.M.; Ibarra, M.R. High-purity cobalt nanostructures grown by focused-electron-beam-induc deposition at low current. *Microelectron. Eng.* **2010**, *87*, 1550–1553. [CrossRef]
63. Mulders, J.J.L.; Belova, L.M.; Riazanova, A. Electron beam induced deposition at elevated temperatures: Compositional chang and purity improvement. *Nanotechnology* **2011**, *22*, 055302. [CrossRef] [PubMed]

4. Roberts, N.A.; Gonzalez, C.M.; Fowlkes, J.D.; Rack, P.D. Enhanced by-product desorption via laser assisted electron beam induced deposition of W(CO)6 with improved conductivity and resolution. *Nanotechnology* **2013**, *24*, 415301. [CrossRef] [PubMed]

5. Roberts, N.A.; Fowlkes, J.D.; Magel, G.A.; Rack, P.D. Enhanced material purity and resolution via synchronized laser assisted electron beam induced deposition of platinum. *Nanoscale* **2013**, *5*, 408–415. [CrossRef]

6. Frabboni, S.; Gazzadi, G.C.; Felisari, L.; Spessot, A. Fabrication by electron beam induced deposition and transmission electron microscopic characterization of sub-10-nm freestanding Pt nanowires. *Appl. Phys. Lett.* **2006**, *88*, 213116. [CrossRef]

7. Plank, H.; Kothleitner, G.; Hofer, F. Optimization of postgrowth electron-beam curing for focused electron-beam-induced Pt deposits. *J. Vac. Sci. Technol. B* **2011**, *29*, 051801. [CrossRef]

8. Gazzadi, G.C.; Frabboni, S. Structural transitions in electron beam deposited Co–carbonyl suspended nanowires at high electrical current densities. *Beilstein J. Nanotechnol.* **2015**, *6*, 1298–1305. [CrossRef]

9. Henry, M.R.; Kim, S.; Fedorov, A.G. High Purity Tungsten Nanostructures via Focused Electron Beam Induced Deposition with Carrier Gas Assisted Supersonic Jet Delivery of Organometallic Precursors. *J. Phys. Chem. C* **2016**, *120*, 10584–10590. [CrossRef]

10. Klein, K.L.; Randolph, S.J.; Fowlkes, J.D.; Allard, L.F.; Meyer III, H.M.; Simpson, M.L.; Rack, P.D. Single-crystal nanowires grown via electron-beam-induced deposition. *Nanotechnology* **2008**, *19*, 345705. [CrossRef]

11. Barth, S.; Huth, M.; Jungwirth, F. Precursors for direct-write nanofabrication with electrons. *J. Mater. Chem. C* **2020**, *8*, 15884–15919. [CrossRef]

12. Friedli, V.; Utke, I. Optimized molecule supply from nozzle-based gas injection systems for focused electron- and ion-beam induced deposition and etching: Simulation and experiment. *J. Phys. D Appl. Phys.* **2009**, *42*, 125305. [CrossRef]

13. Schmied, R.; Fowlkes, J.D.; Winkler, R.; Rack, P.D.; Plank, H. Fundamental edge broadening effects during focused electron beam induced nanosynthesis. *Beilstein J. Nanotechnol.* **2015**, *6*, 462–471. [CrossRef] [PubMed]

14. Pablo-Navarro, J. Development and Optimization of 3D Advanced Functional Magnetic Nanostructures Grown by Focused Electron Beam Induced Deposition. Ph.D. Thesis, Universidad de Zaragoza, Zaragoza, Spain, 2020.

15. Berganza, E.; Bran, C.; Jaafar, M.; Vázquez, M.; Asenjo, A. Domain wall pinning in FeCoCu bamboo-like nanowires. *Sci. Rep.* **2016**, *6*, 29702. [CrossRef] [PubMed]

16. Rodríguez, L.A.; Bran, C.; Reyes, D.; Berganza, E.; Vázquez, M.; Gatel, C.; Snoeck, E.; Asenjo, A. Quantitative Nanoscale Magnetic Study of Isolated Diameter-Modulated FeCoCu Nanowires. *ACS Nano* **2016**, *10*, 9669–9678. [CrossRef] [PubMed]

17. Pablo-Navarro, J.; Sangiao, S.; Magén, C.; de Teresa, J.M. Diameter modulation of 3D nanostructures in focused electron beam induced deposition using local electric fields and beam defocus. *Nanotechnology* **2019**, *30*, 505302. [CrossRef]

18. Manchon, A.; Železný, J.; Miron, I.M.; Jungwirth, T.; Sinova, J.; Thiaville, A.; Garello, K.; Gambardella, P. Current-induced spin-orbit torques in ferromagnetic and antiferromagnetic systems. *Rev. Mod. Phys.* **2019**, *91*, 035004. [CrossRef]

Review

Reversible and Non-Reversible Transformation of Magnetic Structure in Amorphous Microwires

Alexander Chizhik [1,*], Julian Gonzalez [1], Arcady Zhukov [1,2,3], Przemyslaw Gawronski [4],
Mihail Ipatov [1,2], Paula Corte-León [1,2], Juan Mari Blanco [2] and Valentina Zhukova [1,2]

1   Dept. Phys. Mater., Univ. Basque Country, UPV/EHU, 20018 San Sebastian, Spain;
julianmaria.gonzalez@ehu.eus (J.G.); arkadi.joukov@ehu.eus (A.Z.); mihail.ipatov@ehu.eus (M.I.);
paula.corte@ehu.eus (P.C.-L.); valentina.zhukova@ehu.eus (V.Z.)
2   Dept. Appl. Phys., Univ. Basque Country EIG, UPV/EHU, 20018 San Sebastian, Spain;
juanmaria.blanco@ehu.eus
3   IKERBASQUE, Basque Foundation for Science, 48011 Bilbao, Spain
4   Faculty of Physics and Applied Computer Science, AGH Univ. of Science and Technology,
30-059 Krakow, Poland; gawron@newton.fis.agh.edu.pl
*   Correspondence: oleksandr.chyzhyk@ehu.es

Received: 26 June 2020; Accepted: 22 July 2020; Published: 24 July 2020

Abstract: We provide an overview of the tools directed to reversible and irreversible transformations of the magnetic structure of glass-covered microwires. The irreversible tools are the selection of the chemical composition, geometric ratio, and the stress-annealing. For reversible tuning we use the combination of magnetic fields and mechanical stresses. The studies were focused on the giant magnetoimpedance effect and the velocity of the domain walls propagation important for the technological applications. The essential increase of the giant magnetoimpedance effect and the control of the domain wall velocity were achieved as a result of the use of two types of control tools. The performed simulations reflect the real transformation of the helical domain structures experimentally found.

Keywords: soft magnetic materials; amorphous magnetic wires; magnetic domains; magneto-optic Kerr effect; giant magnetoimpedance effect; magnetic anisotropy

---

1. Introduction

Magnetic wires receive permanent interest because of specific magnetic, electric, and magneto-optical properties, which find the realization in wide gamma of electronic devices [1–6].

The importance and novelty of our work is determined basically by the technological application of the microwires. The elucidation of the basic methods for controlling the magnetic structure with the predicted properties is the most promising way for optimizing the operation of magnetic sensors based on wires. We have been dealing with this problem for about 20 years, as evidenced by our first and last work on this topic [5,7].

The nature of glass-coated microwires-circular symmetry, outstanding magnetic properties, and the cheap and accessible fabrication method makes them really attractive from any application's perspectives. The basic concept of many magnetic sensors is tied up with the well-known phenomena of giant magneto-impedance [8–12]. Also, some of these applications like magnetic logic devices, random access memory, integrated circuits, hard disks, and spintronic [13,14] are fare likely rooted to a fast and controllable domain wall motion or propagation throughout magnetic bistability. So far, glass-coated microwires offer such existence of these two functionalities, either GMI or magnetic bistability, in a broad variety of compositions of each of two main families: Co-based or Fe-based alloys, respectively.

Therefore, control of domain wall (DW) dynamics, amplitude of giant magnetoimpedance effect (GMI), and mechanisms of the magnetization reversal, volume or surface, is the key subject of our investigations. Controlled change of the sign and value of the magnetostriction, the reaching of high DW speeds, and switching from axial Barkhausen jump to circular Barkhausen jump and vice versa are the main roots of the improvement of the magnetic properties related with the technological applications [2,15–17].

The approach of our paper is based on the idea to search for methods permitting to control the basic properties of the magnetic microwires. We have focused this paper on these tools influencing on magnetic softness, GMI effect, and single DW velocity in amorphous glass-coated microwires.

Originally, we divided these methods to two groups of reversible and non-reversible changes. These two groups constitute a sequence of actions leading to the expected results. It is evident that the first step of this sequence is the non-reversible change, followed by the finer reversible tuning. Depending on the conditions, different tools could move from one group to another. For example, the torsion stress of the small amplitude causes the reversible changes. The increase of the stress amplitude leads to the un-reversible changes, thereby moving it to the first group. In the same time, the heating causing the annealing process, being lowered in amplitude, leads only to the reversible transformations.

In this work we present the results of the magnetic, magneto-electric, and magneto-optical studies of a series of glass-covered microwires. The studies have been performed in the presence of the different tools induced with reversible and ir-reversible transformation of the magnetic structure. Also, in the frame of the analysis of the experimental results we present the results of the micromagnetic simulations demonstrating the influence of the electric current on the magnetic structure of the microwires.

All of the presented data are our proper experimental results and proper results of the simulations.

2. Materials and Methods

To perform the investigation, we have selected six samples from our wide collection of the glass-covered microwires. The selection of each sample was determined by the concrete task related to the studies of reversible and in-reversible transformations of the magnetic structure. In each of the selected samples, these effects manifested themselves most clearly.

There are the samples which were studied:

- №1 $Fe_{71.7}B_{13.4}Si_{11}Nb_3Ni_{0.9}$, a metallic nucleus diameter of d = 103 μm and a total diameter including the glass coating of D = 158 μm,
- №2 $Co_{41.7}Fe_{36.4}Si_{10.1}B_{11.8}$, diameter of metallic nucleus d = 13.6 μm and total diameter D = 24.6 μm,
- №3 $Fe_{3.6}Co_{69.2}Ni_1B_{12.5}Si_{11}Mo_{1.5}C_{1.2}$, d = 22.8 μm, D = 23.2 μm, annealed at T_{ann} = 300 °C for t_{ann} = 1 h,
- №4 $Co_{67.05}Fe_{3.84}Ni_{1.44}B_{11.33}Mo_{1.69}Si_{14.47}$, d = 21.4 μm, D = 26.2 μm,
- №5 $Co_{66.87}Fe_{3.66}Ni_{2.14}Si_{11.47}B_{13.36}Mo_{1.52}C_{0.98}$ with different diameters of metallic nucleus and total diameters,
- №6 $(Co_{1-x}Mn_x)_{75}Si_{10}B_{15}$ (x = 0.08, 0.09, and 0.10).

The Taylor-Ulitovsky method (also known as quenching-and-drawing) was used to produce the microwires [18,19].

The production of the glass-coated metal filaments of diameter of about a few microns has been reported by Taylor in 1924 [20]. The method of the fabrication is the following. To produce the microwire the metal is held in a borosilicate glass tube. It is closed at one end of the tube with a diameter of about 10 mm. To soften the glass, the end is heated by a glass flame to a specific temperature. At this temperature, the metal part goes to a liquid state. The glass is drawn down and thin glass capillary containing a metal core produces.

The Taylor and Ulitovsky method was revived by Inoue [21], Chiriac [22], and Larin [23]. The modified production technique was developed, which supplied the composite glass-coated microwires with metallic nucleus of diameter in the range of 1–30 μm and glass covering of 2–10 μm. The key parameters of the modified Taylor and Ulitovsky method were the cooling rate of the metal core, chemical composition, and the stable geometrical ratios along the microwire.

Fabrication process of glass covered microwires causes the fast and simultaneous solidification of a composite structure, which consists of magnetic metallic nucleus and glass quenched from the molten alloy. Essential differences in thermal expansion of the glass and the metal lead to large internal stresses. The magnetic structure of an amorphous glass-coated microwire, in the absence of magnetocrystalline anisotropy, is determined basically by the magnetoelastic coupling energy between the internal stresses and spontaneous magnetization.

The stress induced at the fabrication determines the source of anisotropies. The annealing causes a partial relaxing of the internal stresses induced by the production technique, and gives rise to the induced anisotropy. The controlling of the anisotropy is important for technological applications.

Here we present the influence of the annealing with an example of the microwire №3. We consider the annealing as a tool that causes the ir-reversible transformation. The structure of the as-prepared and annealed microwires was recognized by X-ray diffraction [24–27].

In our study we have used the following experimental methods:

Flux-metric method was used for hysteresis loops measurements [28]. The magnetic field was produced by a solenoid. The studied sample was placed in a pick-up coil. The normalized magnetization M/M_0 on external field has been obtained. M is the magnetic moment in some moment; $M0$ is the magnetic moment at the maximum value of the external field Hm. The velocity of DW motion has been measured by Sixtus-Tonks method [29,30].

Originally the Sixtus and Tonks technique consisted of the following elements [31]: a primary coil, which was produced a homogenous-enough magnetic field to induce the propagation of DW wall, short coil induced the nucleation of the DW and two picked up coils. The DW propagation induces the electromotive force in the picked-up coils, which are fixed in the oscilloscope. It was found that in some experimental configurations, the domain wall can nucleate in the middle of the microwire on defects. To avoid mistakes, in the modified set-up we used a system of three pick-up coils [32–35]. The microwire of 10 cm length is placed coaxially inside of the primary coil. To initiate the DW propagation from the determined end of the wire, one end of the sample was placed outside the magnetization solenoid. The distances between coils were 30 mm.

DW velocity was measured as:

$$v = l/\Delta t \tag{1}$$

where time interval Δt is the time difference between the induced peaks and l is the space interval between pick-up coils. We can determine the DW velocity sequentially between 3 pick-up coils.

The impedance Z was recalculated from the reflection coefficient S_{11}, which was obtained using network analyzer:

$$Z = Z_0(1 + S_{11})/(1 - S_{11}) \tag{2}$$

where Z_0 is the impedance of the coaxial line. The GMI ratio $\Delta Z/Z$ was determined as:

$$\Delta Z/Z = [Z(H) - Z(H_{max})]/Z(H_{max}) \tag{3}$$

where H_{max} is the maximum value of the external DC field.

The images of surface magnetic domain structure were obtained by magneto-optical Kerr effect (MOKE) polarizing microscopy (longitudinal geometry). A combination of crossed axial and circular magnetic fields and external mechanical stresses was applied [36]. The torsion stress τ of two opposite signs and limited range of amplitude was applied to the microwires. One of the wire ends was fixed. The other end was rotary stressed to apply the torsion with different angle α and to change the internal stress distribution (Figure 1). The limits of the stress at which the mechanical failure could appear was not achieved.

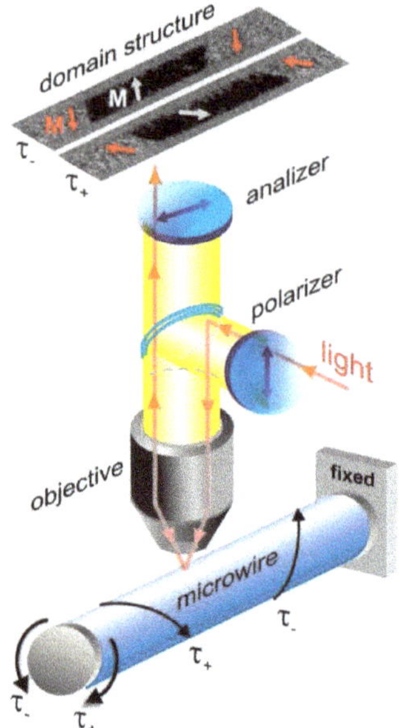

Figure 1. Schematic picture of MOKE geometry and application of torsion stress t$_+$ and t$_-$. Reprinted with permission from [36].

3. Results and Discussions

3.1. Reversible Transformation

Torsion stress and electric current-induced transformations in the sample №1 have been studied using the MOKE technique.

Figures 2 and 3 show the images of magnetic surface domains obtained in the presence of the torsion stress and circular magnetic field. The observed transformations were reversible—when the influence of stress adds circular field was removed, the system returned to its original position.

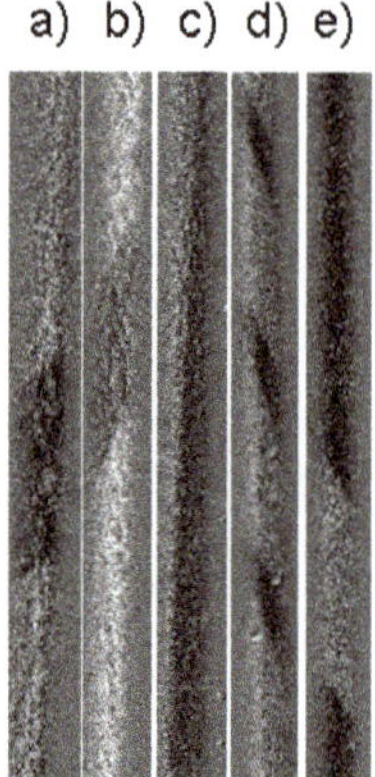

Figure 2. MOKE images of surface magnetic domains. Torsion strain: (**a**) $-8\,\pi\,\mathrm{radm}^{-1}$, (**b**) $-2\,\pi\,\mathrm{radm}^{-1}$, (**c**) 0, (**d**) $2\,\pi\,\mathrm{radm}^{-1}$, (**e**) $8\pi\,\mathrm{radm}^{-1}$. Reprinted with permission from [15].

Figure 2 presents the key details of the effect of the rotation of domain walls induced by torsion stress. There are two types of the helical structure—spiral and elliptical. It was found that the spiral domain structure exists on narrow enough interval of torsion (smaller then 2 πradm^{-1}). For the higher value of torsion, the elliptical structure exists. The angle of the DW inclination changes with the stress as in the case of spiral and in the case of elliptical structures. The DW with longitudinal orientation exists in the presence of the stress of about 0 πradm^{-1} (Figure 2c). Formally, this structure belongs to the spiral type of the structures.

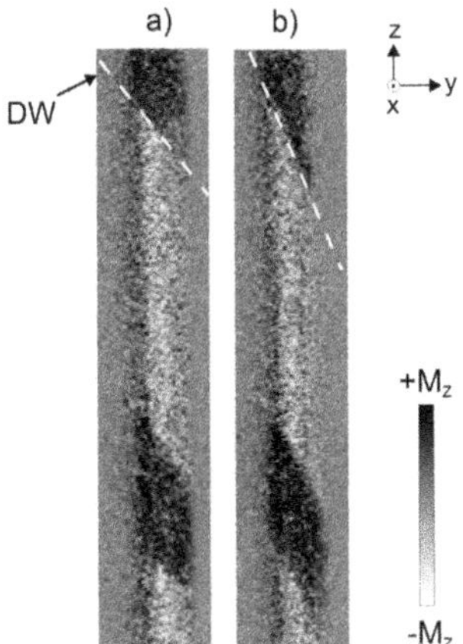

Figure 3. MOKE images of surface magnetic domains. Torsion strain 2πradm^{-1}. Amplitude of circular magnetic field with amplitude: (**a**) 0, (**b**) 7 A/m. White line shows the inclination of domain walls. Reprinted with permission from [15].

Together with DW rotation the torsion stress causes also the transitions between two types of domain structures: spiral and elliptic. The phase transition is reversible for the torsion smaller than 40 πradm^{-1}. This value is the limit of the reversibility. In other words, when this value is exceeded during the experiment, the structure does not return to the spiral structure and the phase transition is not observed.

The circular magnetic field is another tool induced by the reversible transformations and in some experimental situations could be considered as a tool in opposition to the torsion. Figure 3 demonstrates the influence of the circular field. We start in the presence of the torsion stress of 2 πradm^{-1} (Figure 2a). Figure 3b shows the structure in the additional presence of a circular field. The rotation of DW is observed in explicit form towards the axial direction, i.e., against the stress. In the same time, the DW could not reach the longitudinal direction because the limit value of the circular field is determined by the limit value of the electric current induced in this field. The value of this limit is determined by the effect of Joule heating. Therefore, the combination "stress—circular field" is considered as a tool of the fine reversible control of helical magnetic structure.

The dependencies of DM movement on longitudinal magnetic field have been obtained in sample №1 in the presence of small tension stress of about 20 MPa. They were measured by Sixtus–Tonks method. The change of the character of the DW motion as an answer to the torsion application was fixed. The dependencies of the velocity on the longitudinal field in the presence of positive torsion stress are presented in the Figure 4. The similar effect was observed for the negative values of the torsion.

In the absence of the torsion stress, the dependence of the DW velocity is expected and relatively trivial—it is a direct line (black line in the Figure 4). In the presence of the torsion, the noticeable transformation of the velocity dependence takes place. Beginning from the determined value of the stress (about 20 πradm^{-1}), the clear jump is observed on the velocity dependence. In the presence of relatively small magnetic field, the linear dependence with approximately the same slope were obtained for all values of the stress. The linear character of the velocity dependence is practically maintained after the jump but the value of the jump and correspondingly the velocity depends on the stress (blue, green, and lines). The observed effect was named as "torsion-induced acceleration of domain wall".

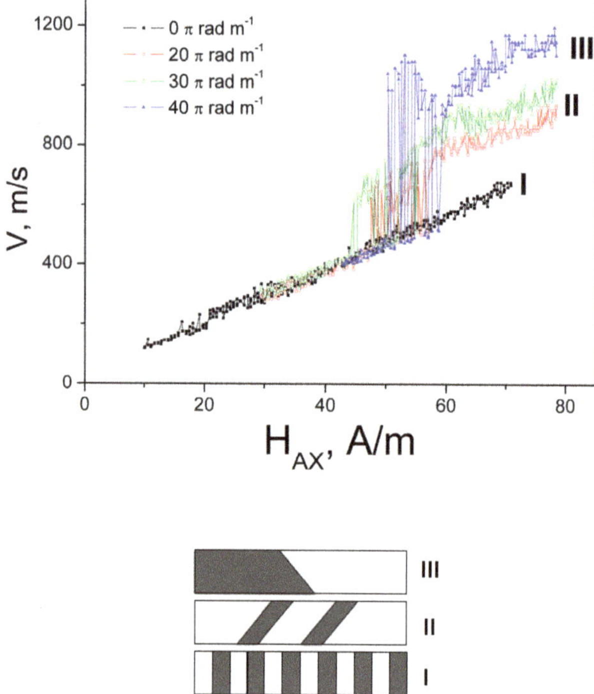

Figure 4. Dependencies of velocity of domain wall on axial magnetic field for different values of torsion stress τ and schematic images demonstrate three types of domain structure. Reprinted with permission from [17].

We have focused on the character of the stress induced of the velocity and the reasons of the dependence of the jump on the stress. It should be noted that such a type of the dependences of DW velocity on the field was not observed earlier. In the narrow interval of the external field (around of 55 A/m) the jumping and falling of the velocity value are observed. These jumping and falling velocities were reversible and random.

The additional MOKE study has been performed to clear this effect. Using the MOKE microscopy we tried to find the main types of the surface domain structures having the relation to the observed stress induced acceleration of DW motion. As a result, we have found 3 different types of domain structure, demonstrated schematically in Figure 4.

The experimental MOKE images of tree different domain structures obtained in the presence of torsion are presented in Figure 5. Earlier it was demonstrated theoretically the existence of different types of helical structures being in energetically close states. They are characterized by close inclination of the magnetization from the axial axis. Now we understand that the induced jumps in the velocity dependence are in direct relation with the reversible transformation between these structures.

Without torsion stress we observed the domains of type *I*, which have the transversal circular direction of the magnetization. The domain of this type is practically un-sensitive to the torsion. The reason of this effect is the strong correlation these domains with the axially magnetized inner core. There is no contradiction with the structures demonstrated earlier because of small tension stress, which was applied during this experiment.

The spiral domain structure (structure of type *II*) exists in a narrow interval of torsion of 20–30 πradm^{-1}. This structure is sensitive on torsion that confirms its independence on the inner core. Spiral structure, along with the elliptic structure, belongs to the helical type of domain structures. We developed some experimental methods, which permit to determine definitely the exact structure of these two observed in the particular experimental conditions.

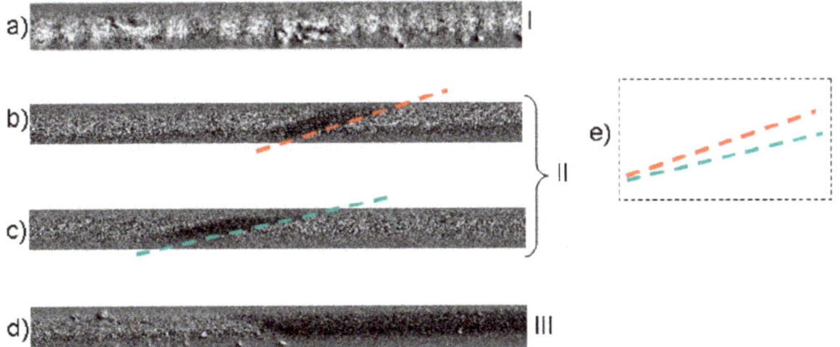

Figure 5. Images of different domain structures obtained in the presence of stress observed for different amplitude of torsion τ: (**a**) 0 πradm^{-1}, (**b**) 20 πradm^{-1}, (**c**) 30 πradm^{-1}, (**d**) 40 πradm^{-1}; (**e**) schematically images of spiral structures. Reprinted with permission from [17].

The increase of the torsion causes elliptic structure (structure of type *II*) forces out the spiral structure from the surface of the microwire. The key difference of these structures is the length of the domain wall. As we know, the domain wall length depends on the angle of the inclination of the domain structure. The elliptic domain wall has a limited length determined by the length of the inclined ellipse. The spiral domain wall has "unlimited" length. Experimentally it is limited only by the length of the studied sample. From the one side, the existence of the elliptic structure is limited by the existence of the stable spiral structure. From the other side, increasing the external stress we reach the natural limit of the angle of the inclination induced by the torsion [37].

The correlation between the domain images presented in Figures 2 and 4 from one side and the velocity dependencies presented in Figure 3 is evident. The stress induced appearance of spiral and elliptical structures recognized in the Figure 3 affects the magnetic field dependence of the DW velocity. Theoretical simulation demonstrates the correlation with the experiment. The induced inclination of the angle of the DW is associated with the DW acceleration.

The change of the domain wall velocity and mobility under the tension stress was studied in microwires № 2. The dependencies of the velocity on the stress with the magnetic field as a parameter are presented in the Figure 6. DW velocity decreases with the increase of the stress.

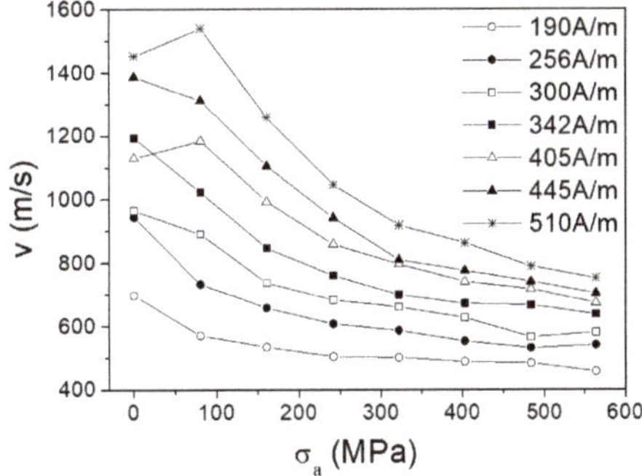

Figure 6. Tension stress dependences of domain wall (DW) velocity in microwires № 2. Reprinted with permission from [6].

The DW mobility S (Figure 7) was extracted from the dependencies presented in Figure 6. The DW mobility decrease with the increase of the tension stress σ_a observed.

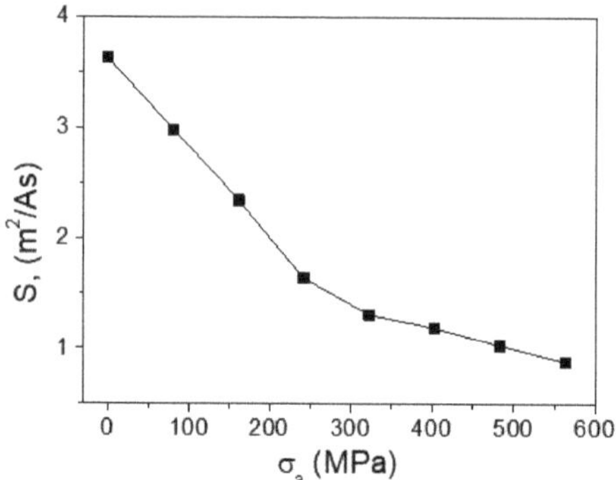

Figure 7. Dependence of DW mobility of microwires № 2 on tension stress. Reprinted with permission from [6].

The velocity v of the DW movement is determined as:

$$v = S(H - H_0) \tag{4}$$

where H_0 is critical propagation field.

For the viscous version of the DW motion the viscosity S is determined as:

$$S = 2\mu_0 M_s/\beta \tag{5}$$

where β is the coefficient of viscous damping, μ_0 is magnetic permeability, M_S is the saturation value of magnetization. In fact, the key parameter which determines the dynamics of DW is the coefficient of damping.

The important component of the damping β is the magnetic relaxation damping, β_r. It is associated with the rotation of electron spins. In turn, it is determined by the parameter α Gilbert damping. The parameter α is proportional inversely to the DW wall width δ_w,

$$\beta_r = \alpha M_s/\gamma\delta_w = M_s(K_{me}/A)^{1/2} \tag{6}$$

where A is the constant of exchange stiffness, K_{me} is the energy of magnetoelastic anisotropy energy, γ is the gyro-magnetic ratio.

This consideration demonstrates the qualitative correlation between domain wall mobility and the magneto-elastic component in the domain wall damping. In other words, when the magnetoelastic anisotropy increases, the domain wall mobility decreases. The achievement of the high mobility of DW requires the decrease of K_{me} constant. To reach it we had to select the composition of the metal nucleus of the microwire with smallest constant of magnetostriction. Another way is the decreasing of internal mechanical stresses by the variation of the geometry of microwire—the relation of the thicknesses of the metallic nucleus and glass covering. The last way is the annealing process causing the relaxation of the internal mechanical stresses.

3.2. Ir-Reversible Transformation

3.2.1. Annealing

The annealing treatment allows the reducing of the coercive field and influences on the DW mobility. We consider this effect as ir-reversible. Taking into account the observed single Barkhausen jump during magnetization reversal, it can be assumed that the single domain wall propagation takes place. Consequently, the measurements of the DW velocity using the Sixtus-Tonks method could be realized in stress-annealed Co-rich microwires.

The series of the microwires № 3 were annealed during 1 h at the temperature of 300 °C in the presence of the tension stress of different values. Figure 8 presents the dependencies of the DW velocity on the external magnetic field for different values of the stress.

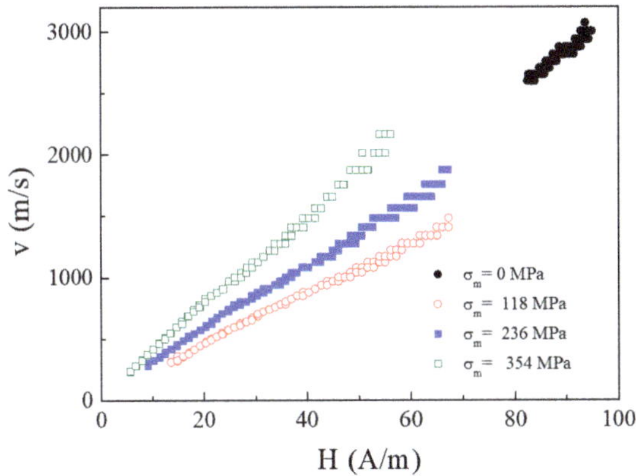

Figure 8. Velocity of DW on axial filed in the presence of different values of the stress applied during annealing. Reprinted with permission from [5].

The microwires annealed without stress, shows the high enough value of the velocity (about 3 km/s) of DW (see Figure 8). The interval of the external field in which the motion of the single domain wall was fixed short (83–93 A/m).

The DW mobility S values are affected by the tension stress σ_m: gradual increase of S values (determined as a slope of $v(H)$ dependencies) is observed upon increase of σ_m (see Figure 8).

The conditions of the annealing treatment influence also on the GMI effect. Figure 9 presents the transformation of the GMI ration field dependencies caused by the tension stress applied during the annealing process. The increase of the GMI ratio is observed with increase of the value of the stress σ_m.

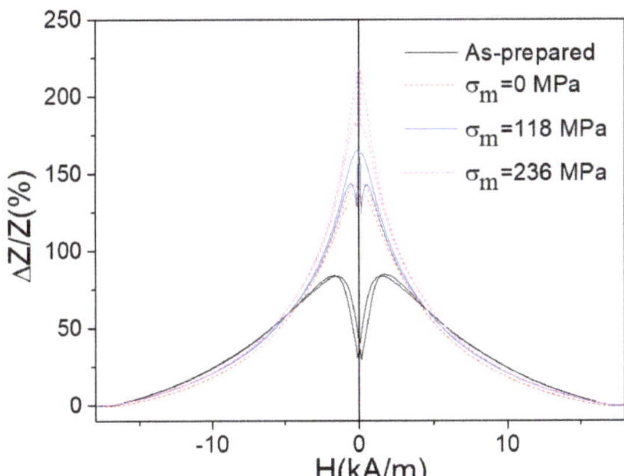

Figure 9. Dependencies of GMI ratio on different value of the external tension stress measured at 200 MHz. Reprinted with permission from [5].

This effect is explained in the frame of the influence of the skin effect on the magneto-electric processes occurring in the magnetic conductor in the presence of the external magnetic field. The bulk hysteresis loops measured by the flux-metric technique provide partial information on the magnetic softness of the whole sample. In the case of the high enough frequency (200 MHz) the GMI ratio characterizes basically the outer layer, which exhibits high circumferential magnetic permeability. In this case, the magnetic softness of the surface layer is more relevant for the GMI effect. The volume of the surface layer is usually much lower than the volume of the inner core and its contribution to the bulk hysteresis loop, generally, is weak.

Here we conclude that the influence of the post-processing on the magnetic processes occurring in the microwires is significant. The variation of the conditions of the annealing is one of the most suitable tools permitted in the predicted optimization of such magnetic properties as domain wall velocity or GMI effect making microwires more suitable for electronic applications.

The field dependencies of the velocity and GMI effect dependencies obtained in Co-rich microwires correlate because the stress applied during the annealing realizes in the stress-induced anisotropy. The stress induced transverse magnetic anisotropy causes the highest GMI effect observed for stress-annealed Co-rich microwires, which present in the same time the rectangular hysteresis loops and hence fast domain wall propagation. The high DW velocity is the result of the fast magnetization switching in the inner axially magnetized single domain.

3.2.2. Geometric Ratio

Magnetic properties of microwires, in particular magnetic anisotropy, are originated basically by the magnetoelastic contribution. In turn, it is determined by the composition of the metal part and by the distribution of the stresses [38–41]. Because of the glass–metal structure of glass-covered microwires, the internal stresses value is determined by the ratio (metallic nucleus diameter d)/(total diameter D) = ϱ. Because of the difference in thermal expansion of metal and glass, the stresses appear during the solidification process, which takes place in the Taylor-Ulitovsky method of microwire production.

The external field dependencies of the DW velocity have been measured in the sample №2 for different values of ϱ (Figure 10). The different slopes on the velocity dependences are observed for different ϱ ratio. As we can see, velocity decreases with the decrease of ϱ ratio and in turn with the increase of the internal stresses.

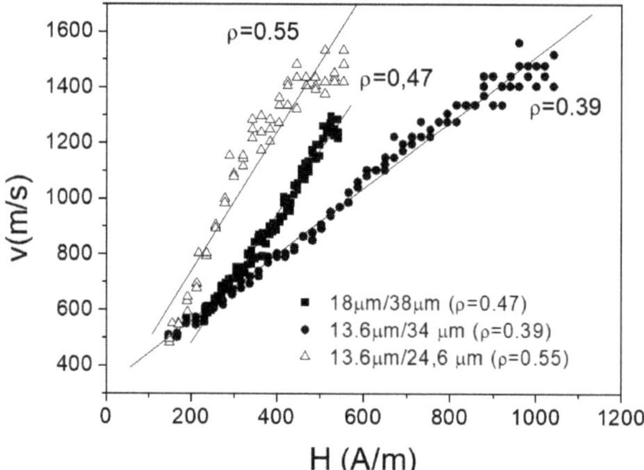

Figure 10. Magnetic field dependences of DW velocity for sample №2 microwires with various ϱ ratio. Reprinted with permission from [6].

Therefore, the velocity is almost two times higher for the highest ϱ ratio: v is about 1500 m/s for ϱ = 0.55 and V is about 900 m/s for ϱ = 0.39, while the composition is the same. The effect of the magnetoelastic energy has been proved by applying the external stress when we have studied the reversible magnetic transformations.

The inset in Figure 11 shows the decrease of the value of the field of GMI maximum with the increasing of the geometric ratio. The influence of the geometric ratio on GMI effect is attributed to the magnetoelastic anisotropy related with the internal stresses. The internal stresses in glass-coated microwires arising from the difference in the thermal expansion coefficients of solidifying metallic nucleus and the covering glass depend in such a way on the ratio between the glass coating thickness and metallic core diameter.

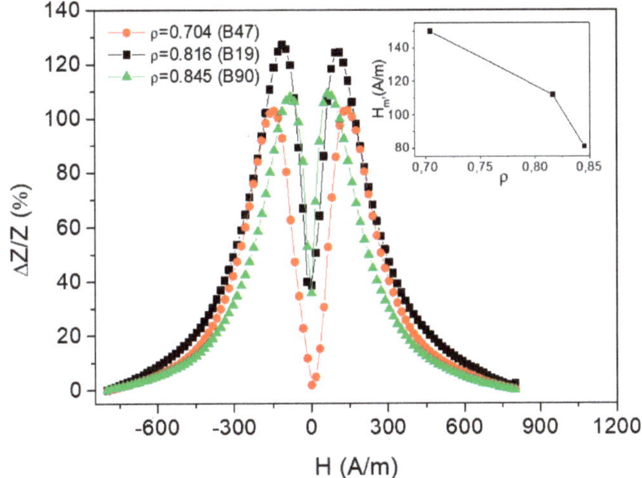

Figure 11. Effect of the sample geometry on magnetic field dependence of giant magnetoimpedance effect (GMI) ratio of microwires №4. Reprinted with permission from [19].

Because the constant of magnetostriction is determined by the chemical composition and has almost zero values in amorphous alloys based on Fe/Co [42–44], the observed dependence should be related to the internal stresses.

We could estimate the amplitude of the internal stresses in glass–metal composition because it is caused by the difference in the coefficients of thermal expansion of the glass covering and the metal nucleus. It should be of about 100–1000 MPa. This value is determined equally by the metallic nucleus radius and the thickness of the glass [6].

Figure 12 demonstrates the GMI effect obtained in a wide range of the frequencies for different diameters of the metallic part of the microwire №5. The highest value of the effect is observed mainly in the range of 100–300 MHz. In the same time, the highest GMI effect takes place and presents at different frequencies for the microwires having the same chemical composition but different geometric parameters. For the wires with d in the range of 9–11.7 μm, the maximum is about 200 MHz, and for the wires with d in the range of 8.5–9 μm the maximum is about 100 MHz.

The change of the position of the maximum H_m is related with the effect that H_m increases with the frequency. For the high frequency and the high applied magnetic field, H_m is approaching to the maximum of the applied field.

At high frequencies, the decreasing of the GMI effect takes place. Therefore, if we have the fixed maximum of external field, the optimum frequency is observed where the highest value of GMI effect takes place. In the same time, the field of GMI maximum is determined by the field of magnetic anisotropy field and we could expect the similar GMI field dependences for similar geometric ratios.

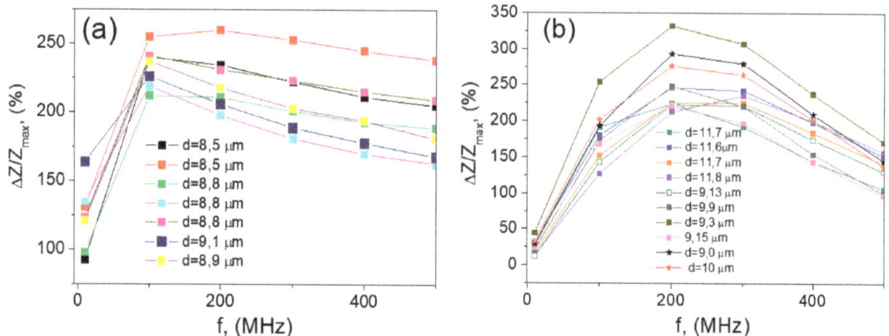

Figure 12. Frequency dependence of microwires №5 with different metallic nucleus diameters. (**a**) metallic nucleus diameters ranging between 8.5 and 9.1 μm; (**b**) metallic nucleus diameters between 9 and 11.7 μm. Reprinted with permission from [19].

3.2.3. Chemical Composition

The chemical composition also has been selected as a tool permitted for the un-reversible establishment of the magnetic properties in microwires. Sample №6 has been chosen to demonstrate the influence of the chemical composition on the magnetization reversal process. In particular the % part of the Mn has been considered as a tool that has been changed.

The magnetic hysteresis obtained for different value of Mn (marked as "x") are presented in Figure 13. The magnetic behavior shown in the Figure 13a (x = 0.08) confirms the transverse magnetic anisotropy. The coercivity for this case is about 5 A/m. The microwire with x = 0.09 demonstrates the intermediate shape of the hysteresis with coercivity of about 12 A/m. Finally, for x = 0.1 the rectangular shape of the hysteresis loop is observed that confirm the effect of magnetic bistability (coercivity about 24 A/m).

The magnetostriction constant λ_S depends on the chemical composition. It changes the sign at Mn concentration value x of 0.09. In spite of low value of λ_S (10^{-7}), the high enough internal stress causes magnetoelastic anisotropy with transversal easy direction for x = 0.08 (λ_S negative) and with longitudinal easy direction for x = 0.1 (λ_S positive).

The hysteresis loops take into account the direction of the easy magnetization axis originated by the magnetoelastic anisotropy. For the case of x = 0.08 axial field induced magnetization rotation associated with the transverse direction of the magnetoelastic anisotropy, while for x = 0.1 the hysteresis with rectangular shape reflects the bistability effect possible only for the longitudinal anisotropy.

This consideration is based on the following relation of the magnetoelastic energy and the magnetostrction constant:

$$K_{me} = 3/2\ \lambda_s \sigma_i, \tag{7}$$

where σ_i are the internal stresses.

To verify this consideration, we have performed the Kerr effect measurements in the same microwire №6 as was studied by the flux-metric technique. The microwire with different concentration of Mn shows the different shape of the transversal and longitudinal MOKE hysteresis loops. The experiments have been performed in circular and axial magnetic fields.

For the case of x = 0.07, the MOKE hysteresis has been obtained in the circular magnetic field (Figure 14a,b). The rectangular shape of hysteresis in Figure 14a means the jump in circular direction. The peaks in longitudinal direction at the same fields confirm the large Barkhausen jump related to the effect of circular bistability. This hysteresis is in agreement with the flux-metric hysteresis presented in Figure 13a.

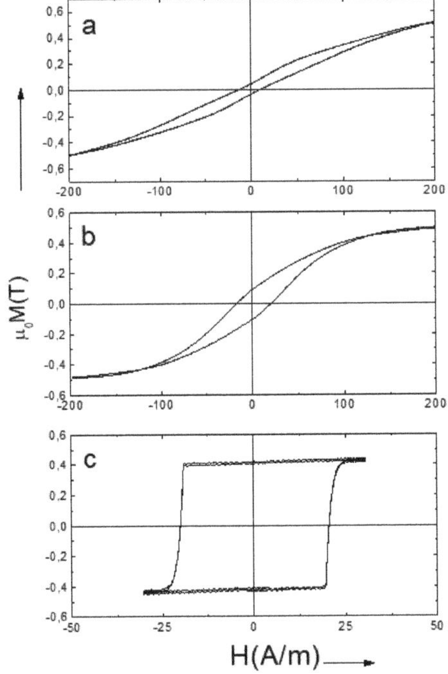

Figure 13. Hysteresis loops the microwire №6. (**a**) x = 0.08, (**b**) x = 0.09, (**c**) x = 0.1. Reprinted with permission from [45].

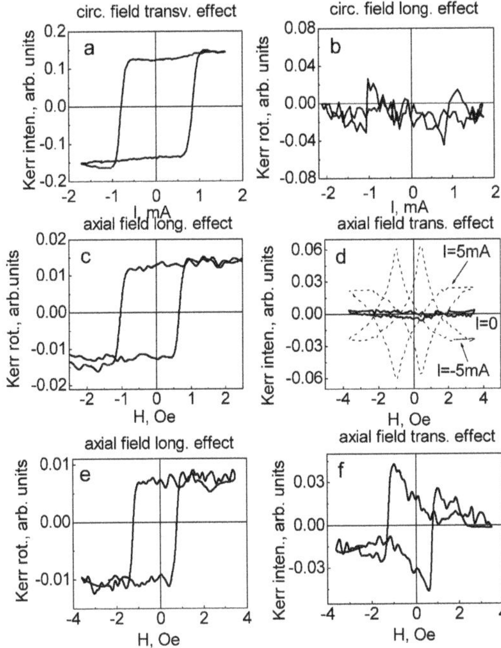

Figure 14. Longitudinal and transverse Kerr effect loops of microwire №6 for different Mn content: (**a,b**) x = 0.07; (**c,d**) x = 0.11; (**e,f**) x = 0.09. Reprinted with permission from [37].

At the same time, for x = 0.11 the rectangular MOKE hysteresis is observed as an axial magnetic field (longitudinal effect, Figure 14c). This shape reflects the axial large Barkhausen jump. Figure 14d (transversal effect) demonstrates a more complicated experiment: the DC circular field of two opposite directions has been added. Without circular field there is no MOKE signal. This effect is related to the magnetization reversal realized as the motion of domain walls between the axial domains. In the presence of the circular field the signal appears because of the field induced inclination of the magnetization from the axial direction.

The microwire with x = 0.09 (Figure 14e,f) is similar to the case of x = 0.11, but some difference should be noted. While the longitudinal MOKE hysteresis in axial field is rectangular, the transversal loop is associated with sharp nucleation of domains and rotation of the magnetization. All of this could be attributed to the effect of helical bistability.

As was mentioned above, the magnetic structure and the magnetization reversal could be associated with sign and the value of the λ_s constant. It was independently determined [46] that λ_s is negative for x < 0.1 and is positive for x > 0.1. The MOKE experiments are in agreement with this independent study taking into account that longitudinal magnetization in the outer shell for x = 0.11 is related to positive λ_s and for x = 0.07 the circular magnetization is related to the negative λ_s.

4. Simulation

For the first time, the calculations of the helical structure in microwires were presented in [47] and were developed in [48]. Here we present the new combination of the hysteresis and images.

There is a procedure of the simulations. The directions and the values of the magnetization were calculated in discrete points of domain structure. Blue and red arrows show the direction of magnetization in the spiral domains.

The blue arrows with maximum intensity correspond to the left direction. This means that the magnetization component is directed along the microwire axis. The red arrows correspond to the right direction. This means that magnetization component is directed along the microwire axis but in the opposite direction then in the plus one case. If the axial magnetization is zero, the color is white and the arrow is perpendicular to the axis of the microwire. Different intensity of the colors means different values of the axial component of the magnetization or different angles of inclination of the arrows in relation to the axis of the microwire.

Figures 15 and 16 present the results of the micromagnetic simulations of the magnetization reversal curves and surface domain images in the microwire. To demonstrate the role of the circular magnetic field, the calculation was performed in the presence of electric current flowing along the microwire to produce the circular field. The amplitude of current does not exceed the critical value related to the overheating and the current is considered as a parameter which induces the reversible changes of the magnetic structure.

The calculations were performed in a magnetic cylinder with discretization of $10 \times 10 \times 5$ nm. In the simulations performed with the help of the mumax program [49], the constant A was taken as 20 pJ/m and the saturation magnetization $\mu_0 M_s$ was equal 0.7T. The magnetoelastic anisotropy is approximated as uni-axial with its spatial dependence related to the experimental stress distribution in a microwire [50]. The magnetoelastic constant was determined using the mentioned above formula (7). The value of magnetostriction constant was 2×10^{-7}.

The simulated magnetic cylinder is subjected to two magnetic fields: variable axial field and circular Oersted field. The external field was changed gradually along the cylinder axis. The initial state was the saturation marked as $m_z/m_s = 1$ and the finish state was the saturation marked as $m_z/m_s = -1$. The control parameter of the simulation was the circular field, which has a distribution in accordance with the Amper's formula.

Figures 15 and 16 present the magnetization reversal curves and the images of the surface domains. The calculations were performed for a current of 1 mA (Figure 15) and 7 mA (Figure 16).

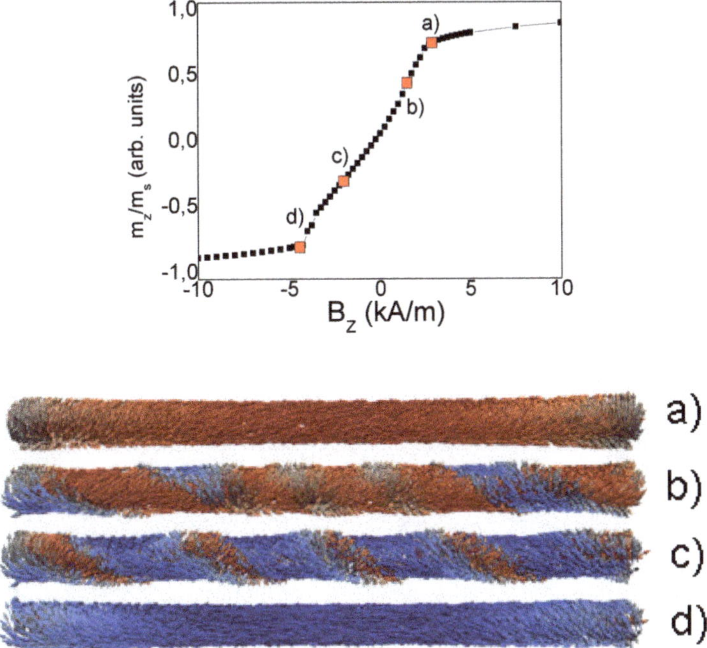

Figure 15. Calculated magnetization reversal curve and the images of spiral domains in cylinder with 1 μm diameter and 15 μm length. Images (**a–d**) correspond to the points (**a–d**) in the hysteresis. Electric current is 1 mA.

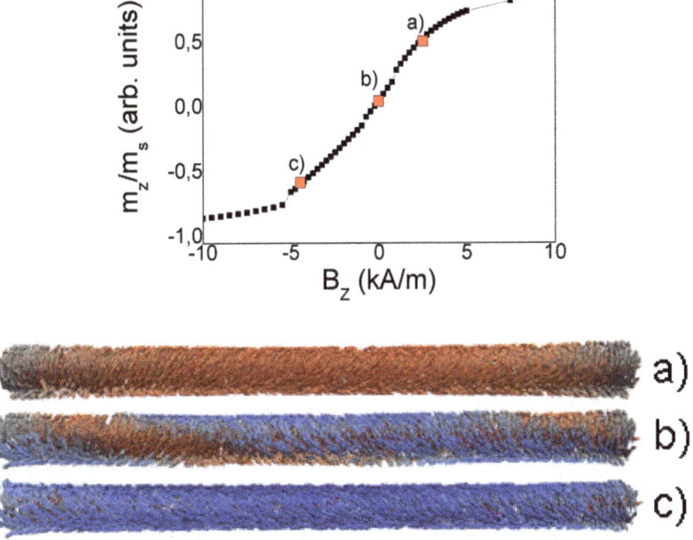

Figure 16. Calculated magnetization reversal curve and the images of spiral domains in cylinder with 1 μm diameter and 15 μm length. Images (**a–c**) correspond to the points (**a–c**) in the hysteresis. Electric current is 7 mA.

The magnetization process begins with the formation of vortices at the ends of the microwire. The vortices, as shown in Figure 8, have the opposite chirality. A decrease in the value of the external applied field causes an inclination of magnetic vectors in the entire volume of the microwire and forms spiral domains. The spiral structure gradually increases from both ends of the microwire towards the center of the microwire (see Figure 15b,c) according to the chirality defined by the vortices. The spiral domain structure is not a formation that occurs only on the surface of the microwire, but also occupies part of the volume inside the microwire. When the spiral is fully formed and extends to the length of the entire microwire, further reducing the value of the applied field, leads to a phenomenon that can be termed spiral propagation. Areas where the magnetic moments are set in accordance with the applied field increase, causing the spiral to move and at the same time unscrew. Eventually, the spiral structure disappears from the surface (see Figure 15d) and its remains in the form of a tube penetrate inside the microwire. Further reduction of the field value also results in setting the magnetic moments inside the microwire according to the field.

At zero circular field, the spiral domain structure consists of two parts with: left- or right-hand chirality. Under applied circular magnetic field, the chirality of the spiral structure becomes uniform in the whole wire, with the direction depended on the direction of the current. If we compare the figures for current of 1 and 7 mA, we will see that an increase of the circular field inclines the magnetic vectors towards the circular direction, perpendicular to the axis of the microwire. It causes a reducing of the density of the spiral structure, (see Figures 15b and 16b), as well as a change of its inclination angle.

The problem of the influence of DW structure on the movable properties of DW has been analyzed in detail in our article [48]. Here, the spiral domain wall demonstrates experimentally the increase of the DW mobility with the increase of the angle of the DW inclination. In turn, the present calculations demonstrate the correlation between the angle of the induced inclination and length of the DW. This result of the simulation could be considered as the most essential result.

5. Conclusions

Here we present the tools that could produce the reversible and irreversible transformations in the glass-covered magnetic microwires. Irreversible transformations are the first step transformations that permit to select and fix the main predicted magnetic and electric properties. From the one side, these properties could not be changed posteriori, but form the other side, they could serve the basis of the stable and repeatable operation of the technological technique. At the second stage we apply the tools caused by the reversible transformations that serve for the fine tuning. This tuning is realized when the conditions of the technique functioning changes.

One of the main tools causing the irreversible transformation is annealing in the presence of stress. The annealing with stress is effective for Co-rich glass-covered microwires. We have observed more than doubled enhancement of GMI ratio. Also, we observed an increase of the DW velocity of about one and a half times in Co-rich microwires. Therefore, the annealing with stress Co-rich microwires demonstrate simultaneously fast domain wall motion and high GMI effect.

The selection of the chemical composition and changing of the thickness of glass covering of the microwire are also considered as tools caused by the irreversible transformation of the magnetic structure that was realized in around two times variation of the DW mobility. Controlling the sign of the magnetostriction, we could obtain the microwires with the longitudinal or transversal large Barkhausen jump.

For reversible tuning, we use the circularly directed external magnetic field and external stresses-tension and torsion. The change of the sign and the amplitude of the torsion induced the change of the direction and the amplitude of the DW inclination. The torsion stress could induce the reversible transition between two types of helical structures: spiral and elliptic which have almost two times different DW mobility in Fe-rich microwires.

In the relation of the tension stress, to get the high domain wall velocity and mobility, the magnetoelastic energy should decrease. This can be achieved by the application of the external stresses and by the control of the distribution of the internal stresses.

The simulations confirm the observed experimental rotation of the helical domain walls induced by the electric current, demonstrating in such a way the correlation between the MOKE experiments and the calculations.

Author Contributions: A.C.: conceptualization, formal analysis, investigation, writing—original draft, writing—review and editing. A.Z.: conceptualization, formal analysis, funding acquisition, investigation, project administration, resources, supervision, validation, writing—original draft, writing—review and editing. M.I.: data curation, methodology. P.C.-L.: data curation, formal analysis, investigation, methodology, writing—original draft, writing—review and editing. J.G.: writing—original draft, writing—review and editing. P.G.: formal analysis, investigation, methodology. J.M.B.: investigation. V.Z.: data curation, investigation, methodology. All authors have read and agreed to the published version of the manuscript.

Funding: This research was funded by National Science Centre Poland under Grant No. DEC-2016/22/M/ST3/00471, Spanish MCIU under PGC2018-099530-BC31 (MCIU/AEI/FEDER, UE), the Government of the Basque Country under PIBA 2018-44 projects. The authors thank for technical and human support provided by SGIker of UPV/EHU and European funding (ERDF and ESF). The research of P.G. was supported in part by PL-Grid Infrastructure.

Conflicts of Interest: The authors declare no conflict of interest.

References

1. Jiles, D. Recent advances and future directions in magnetic materials. *Acta Mater.* **2003**, *51*, 5907–5939. [CrossRef]
2. Vázquez, M.; Chiriac, H.; Zhukov, A.; Panina, L.; Uchiyama, T. On the state-of-the-art in magnetic microwires and expected trends for scientific and technological studies. *Phys. Status Solid A* **2011**, *208*, 493–501. [CrossRef]
3. Mohri, K.; Humphrey, F.; Kawashima, K.; Kimura, K.; Mizutani, M. Large barkhausen and matteucci effects in FeCoSiB, FeCrSiB, and FeNiSiB amorphous wires. *IEEE Trans. Magn.* **1990**, *26*, 1789–1791. [CrossRef]
4. Panina, L.V.; Mohri, K. Magneto-impedance effect in amorphous wires. *Appl. Phys. Lett.* **1994**, *65*, 1189–1191. [CrossRef]
5. Zhukov, A.; Ipatov, M.; Corte-León, P.; Gonzalez-Legarreta, L.; Blanco, J.; Zhukova, V. Soft magnetic microwires for sensor applications. *J. Magn. Magn. Mater.* **2020**, *498*, 166180. [CrossRef]
6. Zhukov, A.; Blanco, J.M.; Ipatov, M.; Zhukova, V. Magnetoelastic contribution in domain wall propagation of micrometric wires. *J. Nanosci. Nanotechnol.* **2012**, *12*, 7582–7586. [CrossRef] [PubMed]
7. Zhukova, V.; Chizhik, A.; Zhukov, A.; Torcunov, A.; Larin, V.; Gonzalez, J. Optimization of gaint magnetoimpedance in Co-rich amorphous microwires. *IEEE Trans. Mag.* **2002**, *38*, 3090–3092. [CrossRef]
8. Karnaushenko, D.; Karnaushenko, D.D.; Makarov, D.; Baunack, S.; Schafer, R.; Schmidt, O.G. Self-assembled on-chip-integrated giant magneto-impedance sensorics. *Adv. Mater.* **2015**, *27*, 6582–6589. [CrossRef]
9. Beach, R.S.; Berkowitz, A.E. Giant magnetic field dependent impedance of amorphous FeCoSiB wire. *Appl. Phys. Lett.* **1994**, *64*, 3652. [CrossRef]
10. Knobel, M.; Vazquez, M.; Kraus, L. Giant magnetoimpedance. In *Handbook of Magnetic Materials*; Bruck, E., Ed.; Elsevier: Amsterdam, The Netherlands, 2003; Volume 15, pp. 497–563.
11. Knobel, M.; Pirota, K.R. Giant magnetoimpedance: Concepts and recent progress. *J. Magn. Magn. Mater.* **2002**, *242–245*, 33. [CrossRef]
12. Phan, M.H.; Peng, H.X. Giant magnetoimedance materials: Fundamentals and applications. *Prog. Mater. Sci.* **2008**, *53*, 323. [CrossRef]
13. Parkin, S.S.P.; Hayashi, M.; Thomas, L. Magnetic domain-wall racetrack memory. *Science* **2008**, *320*, 190–194. [CrossRef] [PubMed]
14. Allwood, D.A.; Xiong, G.; Faulkner, C.C.; Atkinson, D.; Petit, D.; Cowburn, R.P. Magnetic domain-wall logic. *Science* **2005**, *309*, 1688–1692. [CrossRef] [PubMed]
15. Ekstrom, P.A.; Zhukov, A. Spatial structure of the head-to-head propagating domain wall in glass-covered FeSiB microwire. *J. Phys. D: Appl. Phys.* **2010**, *43*, 205001. [CrossRef]
16. Chizhik, A.; Gonzalez, J.; Zhukov, A.; Gawroński, P.; Stupakiewicz, A. Fine tuning of domain helical structure in magnetic microwires. *J. Magn. Magn. Mater.* **2020**, *497*, 166019. [CrossRef]
17. Chizhik, A.; Zhukov, A.; Corte-Leon, P.; Blanco, J.M.; Gonzalez, J.; Gawroński, P. Torsion induced acceleration of domain wall motion in magnetic microwires. *J. Magn. Magn. Mater.* **2019**, *489*, 165420. [CrossRef]

18. Zhukov, A.; Ipatov, M.; Zhukova, V. Advances in giant magnetoimpedance of materials. In *Handbook of Magnetic Materials*; Elsevier BV: Amsterdam, The Netherlands, 2015; Volume 24, pp. 139–236.
19. Zhukov, A.; Ipatov, M.; Churyukanova, M.; Kaloshkin, S.; Zhukova, V. Giant magnetoimpedance in thin amorphous wires: From manipulation of magnetic field dependence to industrial applications. *J. Alloys Compd.* **2014**, *586*, S279–S286. [CrossRef]
20. Taylor, G.F. A method of drawing metallic filaments and a discussion of their properties and uses. *Phys. Rev.* **1924**, *23*, 655–660. [CrossRef]
21. Inoue, A.; Masumoto, T. Properties and applications of amorphous alloy wire. *Engng. Mater.* **1982**, *5*, 47.
22. Chiriac, H.; Óvári, T. Amorphous glass-covered magnetic wires: Preparation, properties, applications. *Prog. Mater. Sci.* **1996**, *40*, 333–407. [CrossRef]
23. Larin, V.; Torcunov, A.; Zhukov, A.; González, J.; Vazquez, M.; Panina, L. Preparation and properties of glass-coated microwires. *J. Magn. Magn. Mater.* **2002**, *249*, 39–45. [CrossRef]
24. Zhukova, V.; Cobeño, A.; Zhukov, A.; Blanco, J.; Larin, V.; González, J. Coercivity of glass-coated $Fe_{73.4-x}Cu_1Nb_{3.1}Si_{13.4+x}B_{9.1}$ ($0 \leq x \leq 1.6$) microwires. *Nanostruct. Mater.* **1999**, *11*, 1319–1327. [CrossRef]
25. Arcas, J.; Gómez-Polo, C.; Zhukov, A.; Vazquez, M.; Larin, V.; Hernando, A. Magnetic properties of amorphous and devitrified FeSiBCuNb glass-coated microwires. *Nanostruct. Mater.* **1996**, *7*, 823–834. [CrossRef]
26. Evstigneeva, S.; Morchenko, A.T.; Trukhanov, A.; Panina, L.; Larin, V.; Volodina, N.; Yudanov, N.; Nematov, M.; Hashim, H.; Ahmad, H. Structural and magnetic anisotropy of directionally-crystallized ferromagnetic microwires. *EPJ Web Conf.* **2018**, *185*, 04022. [CrossRef]
27. Del Val, J.; González, J.; Zukhov, A. Structural and magnetic anisotropy of directionally-crystallized ferromagnetic microwires. *Mater. Sci. Eng. A* **2004**, 679–682. [CrossRef]
28. Zhukova, V.; Blanco, J.M.; Ipatov, M.; Zhukov, A. Magnetic properties and domain wall propagation in FeNiSiB glass-coated microwires. *J. Appl. Phys.* **2014**, *115*, 17A309. [CrossRef]
29. Zhukova, V.; Blanco, J.M.; Rodionova, V.V.; Ipatov, M.; Zhukov, A. Domain wall propagation in micrometric wires: Limits of single domain wall regime. *J. Appl. Phys.* **2012**, *111*, 07E311. [CrossRef]
30. Zhukova, V.; Blanco, J.M.; Ipatov, M.; Zhukov, A. Effect of transverse magnetic field on domain wall propagation in magnetically bistable glass-coated amorphous microwires. *J. Appl. Phys.* **2009**, *106*, 113914. [CrossRef]
31. Sixtus, K.J.; Tonks, L. Propagation of large barkhausen discontinuities. *Phys. Rev.* **1932**, *42*, 419–435. [CrossRef]
32. Varga, R.; Zhukov, A.; Blanco, J.M.; Ipatov, M.; Zhukova, V.; Gonzalez, J.; Vojtaník, P. Fast magnetic domain wall in magnetic microwires. *Phys. Rev. B* **2006**, *74*, 212405. [CrossRef]
33. Hudak, J.; Blazek, J.; Cverha, A.; Gonda, P.; Varga, R. Improved Sixtus–Tonks method for sensing the domain wall propagation direction. *Sens. Actuators A Phys.* **2009**, *156*, 292–295. [CrossRef]
34. Calle, E.; Jiménez, A.; Vázquez, M.; Del Real, R. Time-resolved velocity of a domain wall in a magnetic microwire. *J. Alloys Compd.* **2018**, *767*, 106–111. [CrossRef]
35. Neagu, M.; Chiriac, H.; Hristoforou, E.; Darie, I.; Vinai, F. Domain wall propagation in Fe-rich glass covered amorphous wires. *J. Magn. Magn. Mater.* **2001**, *226*, 1516–1518. [CrossRef]
36. Chizhik, A.; Zhukova, V.; Zhukov, A.; Gonzalez, J.; Gawroński, P.; Kułakowski, K.; Stupakiewicz, A. Surface magnetic structures induced by mechanical stresses in Co-rich microwires. *J. Alloys Compd.* **2018**, *735*, 1449–1453. [CrossRef]
37. González, J.; Chizhik, A.; Zhukov, A.; Blanco, J. Surface magnetic behavior of nearly zero magnetostrictive Co-rich amorphous microwires. *J. Magn. Magn. Mater.* **2003**, *258*, 177–182. [CrossRef]
38. Zhukova, V.; Usov, N.; Zhukov, A.; González, J. Length effect in a Co-rich amorphous wire. *Phys. Rev. B* **2002**, *65*, 134407. [CrossRef]
39. Squire, P.T.; Atalay, S.; Chiriac, H. Δ effect in amorphous glass covered wires. *IEEE Transactions on Magnetics* **2000**, *36*, 3433. [CrossRef]
40. Antonov, A.S.; Borisov, V.T.; Borisov, O.V.; Prokoshin, A.F.; Usov, N.A. Residual quenching stresses in glass-coated amorphous ferromagnetic microwires. *J. Phys. D Appl. Phys.* **2000**, *33*, 1161–1168. [CrossRef]
41. Chiriac, H.; Óvári, T.-A.; Zhukov, A. Magnetoelastic anisotropy of amorphous microwires. *J. Magn. Magn. Mater.* **2003**, *254*, 469–471. [CrossRef]
42. Zhukova, V.; Ipatov, M.; Zhukov, A. Thin magnetically soft wires for magnetic microsensors. *Sensors* **2009**, *9*, 9216–9240. [CrossRef]

43. Konno, Y.; Mohri, K. Magnetostriction measurements for amorphous wires. *IEEE Trans. Magn.* **1989**, *25*, 3623–3625. [CrossRef]

44. Zhukov, A.; Zhukova, V. *Magnetic Properties and Applications of Ferromagnetic Microwires with Amorphous and Nanocrystalline Structure*; Nova Science Publishers Inc.: Hauppauge, NY, USA, 2009; ISBN 978-1-60741-770-5.

45. Zhukov, A.; Zhukova, V.; Blanco, J.; González, J. Recent research on magnetic properties of glass-coated microwires. *J. Magn. Magn. Mater.* **2005**, *294*, 182–192. [CrossRef]

46. Humphrey, F.B.; Mohri, K.; Yamasaki, J.; Kawamura, H.; Malmhall, R.; Ogasawara, I. *Magnetic Properties of Amorphous Metals*; Hernando, A., Madurga, V., Sanchez-Trujillo, M.C., Vazquez, M., Eds.; Elsevier Science: Amsterdam, The Netherlands, 1987.

47. Chizhik, A.; Zhukov, A.; Gonzalez, J.; Gawroński, P.; Kułakowski, K.; Stupakiewicz, A. Spiral magnetic domain structure in cylindrically-shaped microwires. *Sci. Rep.* **2018**, *8*, 15090. [CrossRef] [PubMed]

48. Chizhik, A.; Gonzalez, J.; Zhukov, A.; Gawronski, P. Study of length of domain walls in cylindrical magnetic microwires. *J. Magn. Magn. Mater.* **2020**, *512*, 167060. [CrossRef]

49. Vansteenkiste, A.; Leliaert, J.; Dvornik, M.; Helsen, M.; Garcia-Sanchez, F.; Van Waeyenberge, B. The design and verification of MuMax3. *AIP Adv.* **2014**, *4*, 107133. [CrossRef]

50. Chiriac, H.; Óvári, T.-A.; Pop, G. Internal stress distribution in glass-covered amorphous magnetic wires. *Phys. Rev. B* **1995**, *52*, 10104–10113. [CrossRef]

 *nanomaterials*

...ticle

...n Indirect Method of Micromagnetic Structure Estimation ...1 Microwires

...liia Alekhina [1,2], Valeria Kolesnikova [2], Vladimir Rodionov [2], Nikolai Andreev [2,3], Larissa Panina [2,3], ...leria Rodionova [2,*] and Nikolai Perov [1,2]

[1] Faculty of Physics, Lomonosov Moscow State University, Leninskie Gory 1-2, 119991 Moscow, Russia; ya.alekhina@physics.msu.ru (I.A.); perov@magn.ru (N.P.)

[2] Institute of Physics, Mathematics & IT, Immanuel Kant Baltic Federal University, Gaidara 6, 236041 Kaliningrad, Russia; vakolesnikovag@gmail.com (V.K.); VLRodionov@kantiana.ru (V.R.); andreevn.misa@gmail.com (N.A.); drlpanina@gmail.com (L.P.)

[3] Institute of New Materials and Nanotechnology, National University of Science and Technology "MISiS", Leninsky Avenue 4, 119049 Moscow, Russia

* Correspondence: vvrodionova@kantiana.ru; Tel.: +7-900-346-8482

Abstract: The tunable magnetic properties of amorphous ferromagnetic glass-coated microwires make them suitable for a wide range of applications. Accurate knowledge of the micromagnetic structure is highly desirable since it affects almost all magnetic properties. To select an appropriate wire-sample for a specific application, a deeper understanding of the magnetization reversal process is required, because it determines the measurable response (such as induced voltage waveform and its spectrum). However, the experimental observation of micromagnetic structure of micro-scale amorphous objects has strict size limitations. In this work we proposed a novel experimental technique for evaluating the microstructural characteristics of glass-coated microwires. The cross-sectional permeability distribution in the sample was obtained from impedance measurements at different frequencies. This distribution enables estimation of the prevailing anisotropy in the local region of the wire cross-section. The results obtained were compared with the findings of magnetostatic measurements and remanent state analysis. The advantages and limitations of the methods were discussed.

Keywords: soft magnetic materials; glass-coated microwires; micromagnetic structure; impedance; magnetic permeability

...ation: Alekhina, I.; Kolesnikova, ...; Rodionov, V.; Andreev, N.; Panina, ...; Rodionova, V.; Perov, N. An ...direct Method of Micromagnetic ...ucture Estimation in Microwires. ...*nomaterials* 2021, 11, 274. ...ps://doi.org/10.3390/ ...no11020274

...eived: 17 December 2020
...cepted: 19 January 2021
...blished: 21 January 2021

...blisher's Note: MDPI stays neutral ...h regard to jurisdictional claims in ...blished maps and institutional affil-...ons.

1. Introduction

Amorphous magnetic materials have been thoroughly studied since the 1970s, mainly for soft magnetic applications at elevated frequencies. Over the past 50 years of research, various methods have been developed for prediction of their properties [1,2] which have provided tremendous technical progress in many areas. Apart from their traditional use in motors due to their low eddy currents losses, other practical applications of amorphous magnetic materials range from magnetic field/stress/temperature sensors to logic, coding and memory systems [3–8]. The variety of these applications is due to the specificity of their magnetic and electrical properties (e.g., fast domain wall propagation, giant magnetoimpedance effect) and their sensitivity to external stimuli, such as magnetic field, mechanical load and temperature [9–12]. In many regards, the ability to precisely control their magnetic structure is the basis for the improvement of related technologies.

The magnetic properties of the sample under investigation depend on its micromagnetic structure, which is influenced by magnetic anisotropy. In amorphous materials, the latter is mostly contributed by magnetoelastic interactions depending on the saturation magnetostriction (both on the magnitude and sign) and the spatial distribution of mechanical stresses arising during production or further processing [2,13,14]. To engineer the

magnetoelastic anisotropy in amorphous magnetic materials, different treatment techniqu
may be applied: annealing, drawing, mechanical processing (see, for example, [15–21
The relaxation or redistribution of the mechanical stresses changes the micromagnetic str
ture, which largely determines the mechanism of magnetization reversal (magnetizatic
rotation, domain wall propagation or magnetization jump). The structural changes reve
themselves in measurable effects such as the shape of hysteresis loops [13], permeabili
and magnetoimpedance behaviors [22].

Despite the great importance of the micromagnetic structure, the use of methoc
for its observation has strict limitations. Recently, a novel magnetic tomography methc
based on X-ray magnetic circular dichroism was proposed [23]. This technique mak
it possible to reconstruct the magnetization distribution over the sample and visuali
complex micromagnetic structures such as Bloch points, but the linear dimensions
the samples should not exceed several microns. Most of the experimental techniqu
applicable to bulk magnetic materials allow one to image the micromagnetic structure
the surface [24–26]. Some evidence of a particular magnetic structure can be deduced frc
analysis of the magnetic response from the entire sample or parts of it, for example, frc
static or dynamic hysteresis loops [15,27,28]. The correct interpretation of such indire
experimental results requires precise control of the experimental conditions as well as
large set of measured patterns typical for a particular type of samples.

Here we consider an alternative indirect method for studying the micromagnet
structure of amorphous wires, which allows one to experimentally inspect the intern
domain structure of the sample. The method is based on the impedance measurement
different frequencies.

The impedance of the cylindrical ferromagnetic conductor depends on the perm
ability with respect to the circular magnetic field (so called, circular permeability) if t
skin effect is essential [29]. If this permeability is non-uniformly distributed over t
cross-section of the wire, the current density distribution averaged over a specific lay
becomes a stepwise function of the averaged permeability. The impedance frequen
dependence thus provides information on the permeability averaged over the surface lay
which corresponds to the current penetration depth (the layer involved in the current ma
flow). For this reason, the impedance measurements at several current frequencies allov
the reconstruction of the permeability distribution over the cross-section of a wire by fittir
the experimental and theoretical data.

In this work, additional techniques were used to deduce the magnetization distributic
over the wire cross-section based on magnetization vs. magnetic field behavior. The resul
were compared to assess the reliability of the approaches and possible errors of the methc

2. Materials and Methods

2.1. Materials

2.1.1. Amorphous Magnetic Microwires

Amorphous magnetic microwires were chosen as an object of the investigatio
Being obtained by rapid solidification from the melt, such materials have a compl
micromagnetic structure. Uneven temperature distribution during quenching leads to tin
dispersed solidification of various parts of the wire, which causes non-uniform distributic
of mechanical stresses. Moreover, for glass-covered microwires the difference in the therm
expansion coefficients of glass and metal induces additional stresses in the wire [30,3
Combined with magnetostriction, the stress distribution defines the local directions of t
easy and hard anisotropy axes and the magnitudes of magnetoelastic anisotropy.

The theoretical consideration of the internal stress distribution in amorphous m
crowires based on the quenching process draws a two-region image of the metallic part
the wire: the internal core domain with an axially directed magnetization and the shell wi
a radially or circumferentially directed magnetization, depending on the magnetostrictic
coefficient sign [32,33]. In addition, a more complex magnetic structure can form in the wi
for example, with a helical magnetization [34,35]. The closure domains arising to minimi

the magnetostatic energy should also be taken into account, in particular, for short-length microwires [27,36]. The volume of the inner core domain significantly affects the static and dynamic magnetic properties of microwires: the switching field, the coercivity, the squareness ratio, the range of magnetic fields of single domain wall propagation, and the domain wall mobility. The magnetization distribution also affects the impedance behavior including the change in the shape of impedance vs. magnetic field plots with increasing frequency. Since impedance is a function of the circular permeability, its distribution over the cross-section of the wire has to be taken into account.

2.1.2. Investigated Samples of Amorphous Magnetic Microwires

The amorphous microwires of $Co_{70}Fe_4B_{13}Si_{11}Cr_2$ alloy with near-zero magnetostriction [37,38] produced by Tailor-Ulitovsky technique [39] with different metal cross-section dimensions from 6.4 to 28 µm (6.4, 8, 8.5, 10, 22 and 28 µm) were investigated. The glass shell had the thicknesses of 2.5, 2.8, 2, 2.3, 1, and 3 µm, respectively. The conductivity of amorphous alloy was 8.3×10^5 S. The wire-pieces for measurements were cut by a scalpel.

Because the wire magnetic properties are extremely sensitive to the cross-section size and the phase state (e.g., existence of nanocrystalline clusters which depends on the initial technical parameters [40,41]), we carefully examined the microwire samples using a scanning electron microscope (JSM-6390LV, Jeol, Tokyo, Japan) to determine the relevant dimensions and the transmission electron microscope (JEM-2100 Jeol, Tokyo, Japan, with an accelerating voltage of 200 kV) to detect any presence of nanocrystalline areas. The SEM images are presented in Figure 1a and the TEM micrographs of a thicker wire sample are depicted in Figure 1b,c. A selected area diffraction (SAED) represents an amorphous halo. The dark-field observation in the first diffraction ring also confirms the amorphous structure.

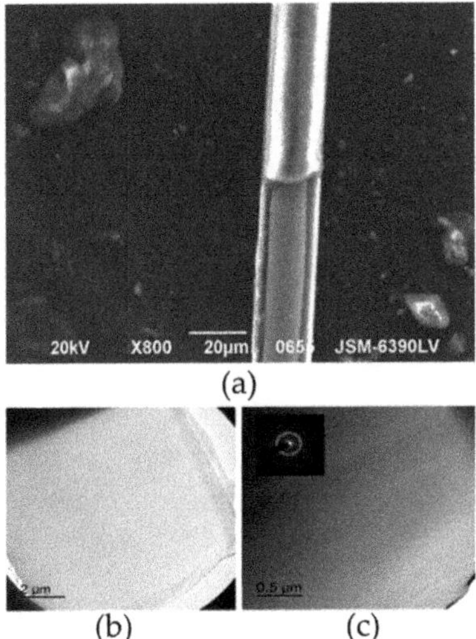

Figure 1. (**a**) SEM and (**b**) TEM images of $Co_{70}Fe_4B_{13}Si_{11}Cr_2$ microwire. (**c**) SAED from the microwire and subsequent dark-field image obtained in the first diffraction ring. The samples demonstrate an amorphous structure.

2.2. Methods

For the microstructure investigations, two basic approaches were used. The first is based on the analysis of the impedance characteristics at different frequencies to reproduce the permeability distribution, and the second utilizes the analysis of hysteresis loops: remanent magnetization and re-magnetization mechanisms.

2.2.1. Impedance of the Wires and Permeability Calculation
Theoretical Background

For moderate frequencies (not very strong skin effect), the impedance of a magnetic wire with a spatially independent permeability tensor is of the form [29]:

$$Z = R_{DC} \frac{ka}{2} \frac{J_0(ka)}{J_1(ka)}$$

$$k = \frac{1-i}{\delta}, \ \delta = \frac{1}{\sqrt{\pi f \sigma \mu_0 \mu_\varphi}},$$

where R_{DC} is the DC resistance, J_0 and J_1 are the Bessel functions of zero and first order respectively, δ is the skin depth, f is the current frequency, σ is the wire conductivity, μ_0 is the vacuum permeability, and μ_φ is the circular permeability, a is the wire radius.

This expression explains the typical shapes of the giant magnetoimpedance (GMI) curves. The magnetic field dependence of the circular permeability, included in Equation (1) as the function argument, is determined by the magnetic anisotropy. A bell-shaped curve of GMI is associated with an axial anisotropy, whilst a circular anisotropy results in GMI plot with two symmetrical peaks appearing at the external field nearly equal to the anisotropy field [22]. In some cases, CoFe-based microwires show asymmetry in the GMI plots. Such behavior is explained in terms of the combination of a helical magnetic anisotropy and circular field produced by the bias current during the measurements [42,43].

Equation (1) for the impedance is obtained considering a linear relationship between AC magnetization and magnetic field under the assumption of a spatially independent permeability tensor. In general, the permeability parameter μ_φ is composed of the components of the permeability tensor. In real samples the permeability may be distributed over the volume, and its dependence on the magnetic field (which is also non-uniformly distributed over the wire cross-section) have to be taken into account. It is assumed, that for a smooth radial dependence of permeability the same expression works if the parameter μ_φ is replaced by an average permeability of the layer related to the skin depth. For the detailed reconstruction of the permeability distribution in complex cases of the dependence $\mu(r, H)$, the calculation method has to be sufficiently modified. A simplified approach makes it possible to deduce some tendencies of the permeability spatial distribution.

In the first approximation, assuming a complex-valued permeability $\mu_\varphi = \mu'_\varphi - i\mu_\varphi''$, the real and imaginary parts can be reconstructed by solving the inverse problem. The minimum position of the function ΔZ:

$$|\Delta Z| = |Z_{exp} - Z_{th}|$$

where Z_{exp} is the measured impedance and Z_{th} is the theoretical value determined by Equation (1), on $\{\mu'_\varphi, \mu_\varphi''\}$ surface corresponds to the permeability value averaged over the wire layer, where the bulk of the current flows. We assume that 70% of the current flow is the main contributor.

The current density in the wire depends on the radius due to the skin-effect. The current integrated over the surface layer of thickness $h = a - r$ equals:

$$I_h = I_0 \left(1 - \frac{r}{a} \frac{J_1(kr)}{J_1(ka)}\right),$$

where I_0 is the total current passing through the wire. Using Equation (3), the current penetration depth $h_{70\%}$ where 70% of the current flow is concentrated, can be estimated. As an example, for a sample with a diameter of 8 μm the current density distribution at various frequencies from 1 to 10 MHz is shown in Figure 2, where the penetration depth and average permeability are also indicated.

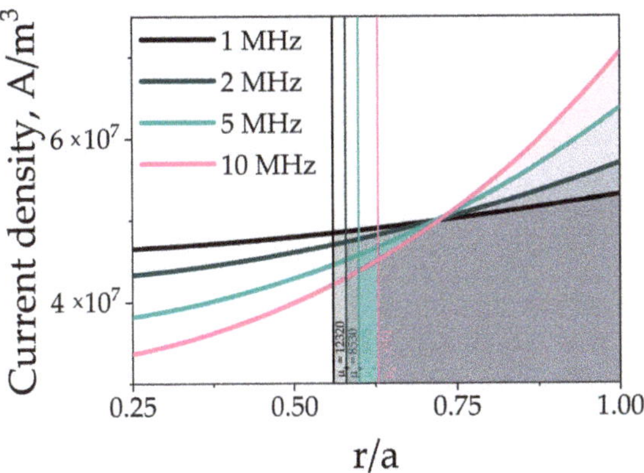

Figure 2. Radial dependence of the current density at frequencies of 1, 2, 5 and 10 MHz. The vertical lines bound the layers through which 70% of the total current flows (shaded areas). The average permeability values for each case are marked with the corresponding color (these values were found by experimental data fitting procedure). For a frequency 1 MHz, the permeability μ_φ is of the order of 10^4, which is typical for low-magnetostriction wires.

The AC current passing through the wire produces an AC circular magnetic field, which causes the AC magnetization. In the presence of a DC magnetic field, the AC magnetization process occurs by the both mechanisms: magnetic moment rotation and domain wall movement. With increasing frequency towards MHz range the main contribution comes from the magnetization rotation as the domain walls are damped [44,45].

Varying the frequency, the changes in $h_{70\%}$ and corresponding μ_φ can be obtained. As measured μ_φ is an average value of permeability over the layer with thickness $h_{70\%}$, knowing the array of μ_φ corresponding to different $h_{70\%}$, one can reconstruct the radial dependence of permeability μ_φ^r, as depicted in Figure 3.

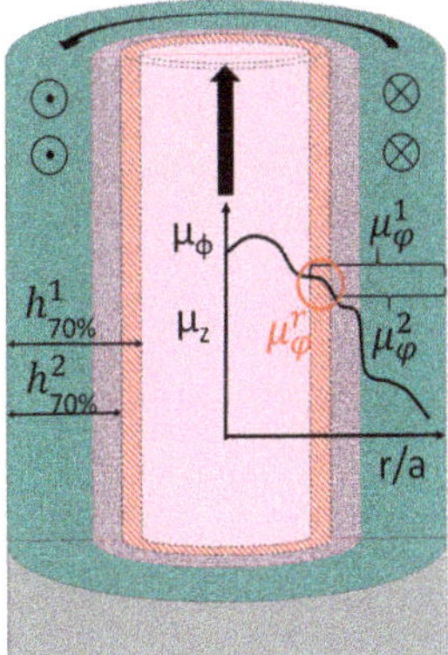

Figure 3. Schematic of the permeability distribution calculation based on averaging the permeabilities over the penetration depths. The values of the average permeability μ_φ^1 and μ_φ^2 correspond to the experimental impedance measured at two frequencies $f(f_1 < f_2)$. The permeability values are used to estimate the corresponding penetration depths $h_{70\%}^1$ and $h_{70\%}^2$. From average values, the permeability μ_φ^r of the differential layer (red) can be interpolated.

If the current penetration depth falls into the axially magnetized region, the average permeability associated with the magnetization rotation increases due to the perpendicular configuration between the DC magnetization and AC magnetic field. We assume that this enhanced permeability corresponds to the axial anisotropy, which also affects the magnetization reversal process and can be revealed in hysteresis experiments. Consequently, analyzing the radial dependence of permeability and identifying abrupt changes in μ_φ, is possible to trace the regions in which the prevailing type of anisotropy changes.

For the permeability calculation the following procedure was used:

(i)   using the experimental data on the wire impedance, the real and imaginary parts permeability were determined from Equations (1) and (2);
(ii) knowing the permeability value, the corresponding current penetration depth every frequency was obtained from Equation (3);
(iii) for each subsequent pair of permeability and penetration depth values the average permeability for the differential layer (red layer in Figure 3) was obtained. The value obtained were presented in the form of a histogram, where the pillar height and width represent the local permeability and the thickness of the differential layer, respectively.

2.2.2. Magnetostatic Measurements

There are several indirect experimental methods for obtaining data for estimating the volume of the axially magnetized core of the microwire based on magnetostatic measurements. Local hysteresis loops along the microwire axis can be obtained by measuring the magnetic flux in short movable pick-up coils. This also allows the magnetization profile be measured in order to assess the contribution of closure domains [15,27,46,47].

When using integrated fluxmetric methods for studying magnetic properties, the closure domains do not contribute to the measured hysteresis loop. The estimation of closure domain volume may be more accurate comparing to that obtained using vibrating sample magnetometry methods (VSM). However, the AC fields are used in the fluxmetric method, which causes the magnetic parameters to vary with a frequency and an instant value of the excitation magnetic field [28].

In the case of using the VSM methods, the closure domains contribute to the measured hysteresis loops. Moreover, in a Vector VSM setup, the two pairs of pick-up coils are located at some fixed distance from each other and the sample is placed between them. This means that different parts over the length of the elongated specimen have different contribution to the magnetic moment projections and the final estimate of the volume of the axially magnetized core is either too large or too small.

In this work, VSM with a modified pick-up coil configuration was used for magnetostatic measurements. We used double half-ring coils having a diameter of 2 cm and 10^4 turns copper wire with a thickness of 30 μm, as schematically shown in Figure 4 by grey line. The proposed pick-up coil arrangement offers a possibility of measuring different projections the magnetic moment keeping the same value of the magnetic flux from the sample [48].

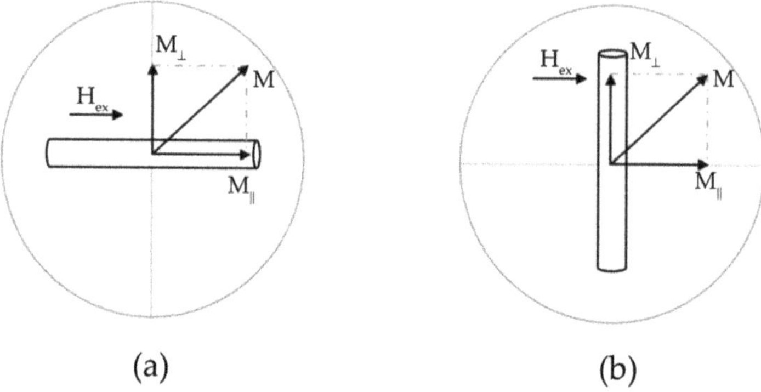

Figure 4. Schematic diagram showing the principal directions of the measured magnetic moment with respect to the applied magnetic field H_{ex}: (**a**) component of the wire magnetic moment along the field H_{ex} as a function of H_{ex}, (**b**) cross-magnetic moment vs. H_{ex}.

The measurements of (a) conventional hysteresis loops, that is, the component of the wire magnetic moment along the magnetizing field H_{ex} as a function of H_{ex} and (b) perpendicular hysteresis loops, that is, the component of the wire magnetic moment perpendicular to H_{ex} vs. H_{ex} (named as cross-magnetic moment) were carried out. The measurement geometry s for both cases is demonstrated in Figure 4: the axial magnetic moment was measured with respect to the magnetic field applied along the wire (a) and perpendicular to the wire axis (b).

Radius estimates of the core domain can be made using two algorithms. The first one is based on the analysis of the remanent magnetization measured along the external field H_{ex} as in (a) case. For this configuration, the remanent magnetization M_r corresponds to the core domain magnetization (with the contribution of closure domains) and the saturation magnetization is a characteristic of a uniformly magnetized wire. Figure 5 shows a typical micromagnetic structure of the metallic part of a glass-coated microwire, from which it can be seen that the squareness ratio of the hysteresis loop $\frac{M_R}{M_S}$ defines the volume fraction V_C of the axially magnetized core with respect to the entire sample volume V_S. Thus, the radius of the axially magnetized core can be estimated.

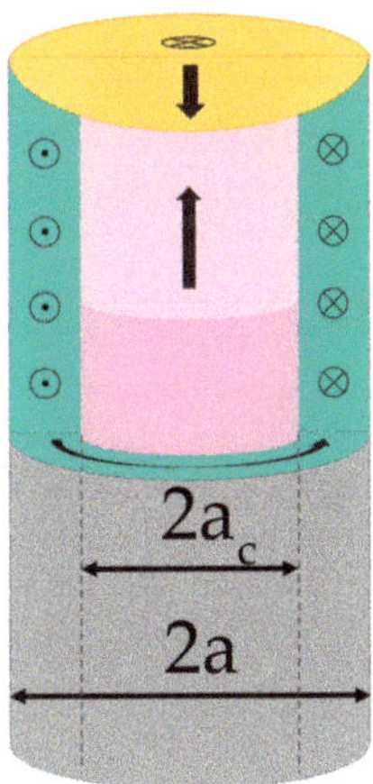

Figure 5. Schematic of the micromagnetic structure of the metallic part with a diameter 2a of a magnetic microwire: axially magnetized core of a diameter 2a$_c$ (pink) and circularly magnetized shell (green). Closure domains are presented by the yellow region.

The second method uses the results of hysteresis loops measurements in both configurations (a) and (b). This requires the comparison of the maximum values of the magnetic moment components measured in parallel (a) and perpendicular (b) orientations. The maximum magnetic moment in case (a) corresponds to M_S. The maximum value of the cross-magnetic moment, M_C, is the magnetic moment of the axially magnetized core only originated from the axial anisotropy. This excludes the closure domains contribution. Similar to the squareness coefficient, their ratio shows the volume fraction of the core:

$$\frac{M_S}{M_C} = \frac{V_S}{V_C} \tag{4}$$

Thus, we can estimate the location of the domain wall between the axially magnetized core and the circumferentially magnetized shell inside the microwire:

$$\frac{V_s}{V_c} = \frac{a^2}{a_c^2} \tag{5}$$

Using Equation (5), the radius of the axially magnetized core can be expressed by:

$$a_c = a \sqrt{\frac{V_c}{V_s}}$$

(6)

2.2.3. Experimental Equipment

The measurements of the impedance Z_{exp} were carried out using a HP4395A network/spectrum/analyzer (Hewlett-Packard, Palo Alto, CA, USA) which was calibrated to get the data of high resolution in MHz frequency range. The measurements were carried out when an alternating current with an amplitude of 2.5 mA kept fixed by the hardware and a frequency of f = 0.5–10 MHz was passing through the sample. The length of the microwire sample was 8 mm.

The calculations using the procedure described in Section 2.2.1 were carried out with the use of MatLab R2015b (MathWorks, Santa Clara, CA, USA) software.

The measurements of magnetic moment were carried out using self-assembled VSM with double-split measuring coils [48]. The magnetic moment resolution was 10^{-3} Am. The range of the applied magnetic field was ± 800 kA/m. The field increment in the low field range was 2.4 A/m. The measured magnetic moments were given in arbitrary units which correlated with the induced voltage in the pick-up coils. The length of microwire samples was 1.5 cm which is sufficient to maintain the original micromagnetic structure for microwires with mentioned diameters and compositions [49].

3. Results and Discussion

3.1. Conventional Hysteresis Loops

The shape of the hysteresis loops is a good indicator of the magnetization mechanism. Figure 6 compares the hysteresis loops of CoFe-based wires with different metallic core diameters measured in (a) geometry (see Figure 4). They all have low coercivity (lower than 5 A/m) typical of good soft magnetic materials. The saturation field does not exceed 40 A/m. The squareness ratio is from 0.02 to 0.45. The parameters are listed in Table 1. The radii of the axially magnetized core were calculated for each sample from the squareness ratio and varied from 0.05 to 0.67 of the total core radius.

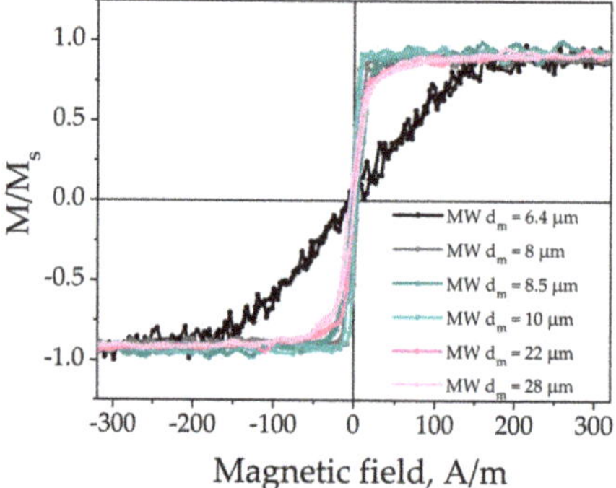

Figure 6. Normalized hysteresis loops of the $Co_{70}Fe_4B_{13}Si_{11}Cr_2$ amorphous glass-coated microwire with different diameters. The shape of hysteresis loops evolves depending on the diameter of the metal core.

Table 1. Magnetic parameters of the $Co_{70}Fe_4B_{13}Si_{11}Cr_2$ amorphous microwires.

Metallic Core Diameter, d_m [μm]	Coercivity, H_C [Oe]	Remanence to Saturation Ratio, M_R/M_S
6.4	0.05	0.05
8	0.05	0.45
8.5	0.05	0.24
10	0.03	0.33
22	0.04	0.35
28	0.003	0.02

d_m is the diameter of the metallic core; H_C is the coercive field; M_R/M_S is the squareness ratio, M_R is the remanent magnetization, M_S is the saturation magnetization.

3.2. Field Dependence of the Cross-Magnetic Moment

Figure 7 shows the plots of the axial magnetic moment vs. the field applied perpendicular to the wire axis (cross magnetic configuration). In the saturation field this component tends to zero since the wire is magnetized in perpendicular direction. The magnetization mechanisms are illustrated in Figure 7 (right column) where M_1 and M_2 are the magnetic moment of the core (with the axial anisotropy) and the periphery, respectively.

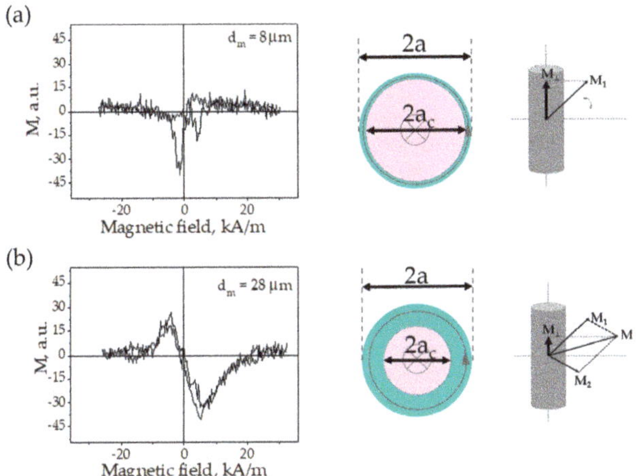

Figure 7. Axial magnetic moment vs. magnetic field applied perpendicular to the wire axis for $d_m = 8$ μm in (**a**) and 28 μm in (**b**).

Figure 7a clearly demonstrates the existence of the domain processes in the region of low magnetic fields resulting in a sharp peak of the cross-magnetic moment. The axial orientation of the magnetization is assisted by the perpendicular field owing to the corresponding components of the internal stress. Figure 7b is consistent with the rotational mechanism of the magnetization. The values of the magnetic moments in Figure 7a,b should not be compared as they are obtained for microwires with different diameters. For further analysis only relative values are of interest. Analyzing the relative data for the two types of magnetization curves, it will be possible to draw conclusions about the distribution of the easy anisotropy axes (either axial or circumferential) within the microwire cross-section.

3.3. Remanent Magnetization State Analysis

Using the results of the two types of measurments (conventional hysteresis loops and cross-magnetization loops), we estimated the volume of the axially magnetized core for all samples. The results are summarized in Table 2. The differences in estimations are clearly seen. The reason is the difference in the magnetization processes of the wire when magnetic field is applied parallel or perpendicular to the wire axis.

Table 2. Estimated radius of axially magnetized core by two methods.

Metallic Core Diameter, d_m [μm]	Core Domain Radius, a_c/a	Core Domain Radius, a_c */a
6.4	~0	0.22
8	0.77	0.67
8.5	0.55	0.49
10	0.54	0.57
22	~0	0.60
28	~0	0.14

d_m is the diameter of the metallic core; a_c is the radius of the axially magnetized core of the microwire metal part; a_c * is the radius of the axially magnetized core of the microwire metallic part calculated on the basis squareness ratio; a is the radius of the metallic part.

Based on the squareness coefficient, we could assume that the microwire with a diameter $d_m = 28$ μm has quite a large axial domain volume comparable with that of thinner wire (0.6 a). On the other hand, the axial magnetic moment tends to zero when the perpendicular field decreases to zero. This indicates that the magnetic moment of the remanent state conventional measurements can be decreased due to the contribution of closure domain. It means, that for samples with large closure domains the core domain radius cannot be accurately found on the basis of the squareness coefficient. The cross-magnetic moment measurements provide more accurate information about the remagnetization mechanism and prevailing anisotropy allowing the magnetization evaluations without the closure domain contribution.

3.4. Magnetoimpedance

GMI dependences for all the samples were analyzed in order to get the information about the prevailing anisotropy type and for further establishments of model working frames (Figure 8). A two-peak GMI curve is inherent of wires with a circular anisotropy. When axial or radial anisotropy prevails, a single-peak dependence is usually observed [2].

For the wires under investigation the shape of the GMI curve evolves with the increase in the metallic core diameter. For the thinnest sample the two-peak plots were observed. When the diameter of the wire increases, the field of the GMI maximum shifts to lower values, and for samples with $d_m = 8.5$ and 10 μm the dependence with a single peak was observed. With further d_m increase, a single peak splits again into two peaks. This demonstrates a non-monotonic change in the domain structure with an increase in the wire diameter, which can be explained by the difference in the stress distribution and the dependence of magnetostriction on stresses.

It was reported, that for amorphous alloys the magnetostriction coefficient depends on the magnitude of the mechanical stresses. For materials with nearly zero magnetostriction it can change the sign under the mechanical influence [50–53]. As the internal stress in the wire depend on the geometrical and manufacture parameters, the wires of the same compositions but prepared at different conditions may have opposite signs of the magnetostriction coefficient and, thus, different types of the domain structure [50,54,55].

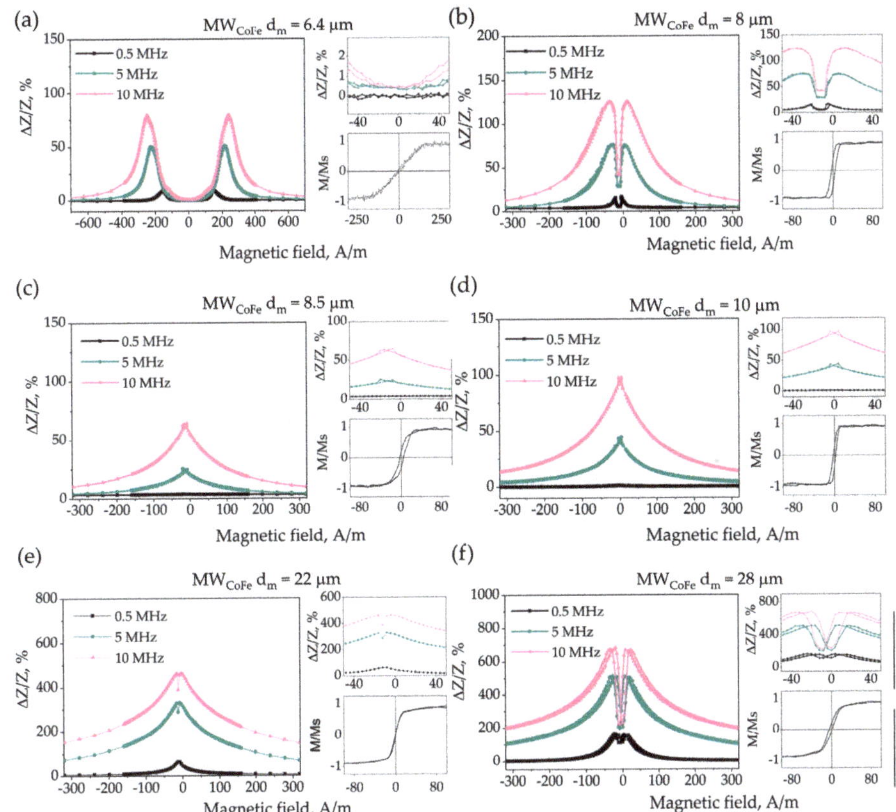

Figure 8. GMI field dependences at different frequencies for microwires with different metallic core diameters from 6.4 to 28 μm (**a**–**f**). Enlarged near-zero field region and hysteresis loops for the samples are included. Evolution of GMI curve shape from double peak (**a**,**b**) to single-peak (**c**,**d**) and back to double-peak (**e**,**f**) with increasing metal core diameter is observed.

The GMI curves for microwires with d_m = 6.4, 8, 22 and 28 μm have two peaks, that is, they have a predominant circular anisotropy. For the thinnest sample (d_m = 6.4 μm) there is a wide plateau around zero field, which is associated with a strong circular anisotropy field [56]. The shape of the hysteresis loop (Figure 8a) and the absence of cross-magnetic moment indicate that the wire has no core in the domain structure. The GMI curve for the sample with d_m = 8 μm also demonstrates two peaks with narrow plateau. Taking into account the shape of cross-magnetic moment curve, it is supposed, that the regions of axial and circular magnetization coexist in this wire. For the samples with d_m = 22 and 28 μm both the cross-magnetic moment loops and GMI curves allow us to conclude, that circular magnetization predominates.

For the samples with d_m = 8.5 and 10 μm a single-peak GMI curve is observed. The comparison of GMI with the results of magnetostatic measurements for these samples showed, that there was a large fraction with an axial magnetization in the wires. Considering the GMI curves and possible stress-dependence of the magnetostriction it is assumed, that internal stresses induced a change in the magnetostriction sign, and the radial magnetization distribution prevails in the shell of these samples.

3.5. Radial Distribution of Magnetic Permeability

Figure 9 demonstrates the radial distribution of circular permeability for all samples. Since the impedance of the wires was measured at MHz frequencies, the main contribution to the permeability comes from the magnetization rotation, while the motion of domain walls is suppressed by eddy currents. This is confirmed by previous studies of the frequency behavior of permeability in microwires of similar composition [45]. However, at zero DC field and circular anisotropy, the AC circular field does not cause magnetization rotation and only the displacements of circular domain walls affect the permeability. The current density distribution at various frequencies from 1 to 10 MHz and corresponding current penetration depths and average permeability values for the sample with 8 μm diameter are presented above in Figure 2.

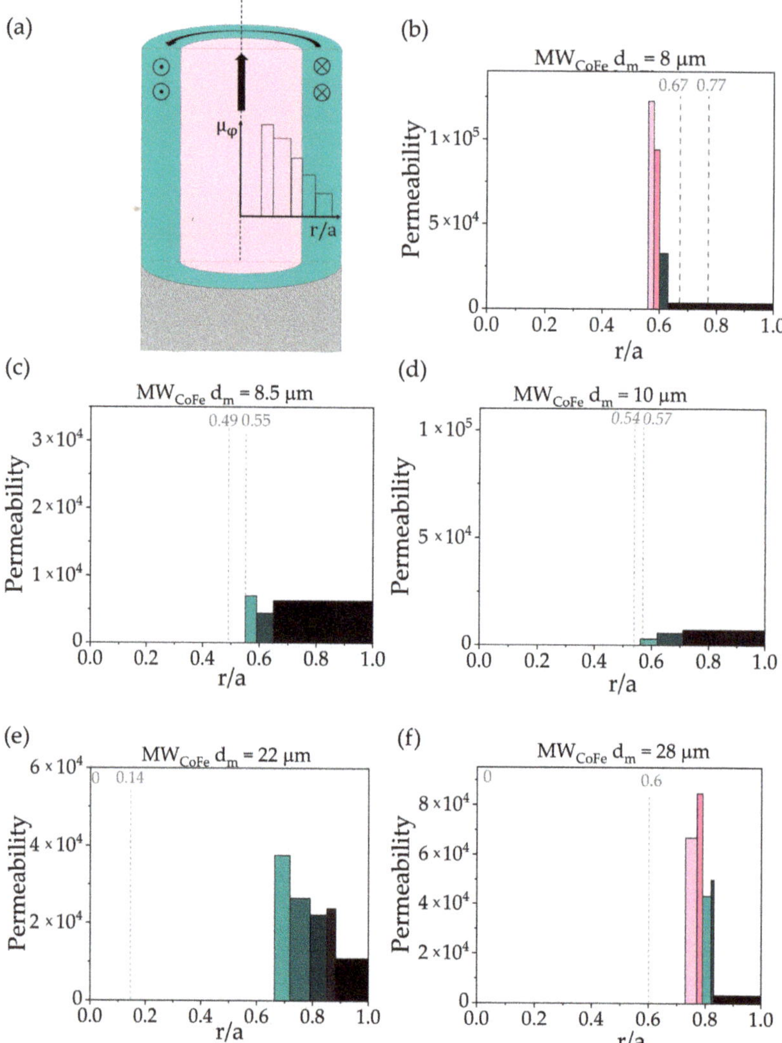

Figure 9. Radial distribution of the permeability of CoFe-based microwires with different diameters. (**a**) Schematic representation of the permeability histogram construction reflecting radial permeability distribution, (**b–f**) permeability distribution histograms for the samples with different metal core diameters. The height of the bar is the local permeability value and the width is the thickness of the corresponding layer. Color is used to contrast the permeabilities of different layers. The permeability calculation was done for zero DC magnetic field.

The results obtained demonstrate, that for the thinnest sample with a diameter of 6.4 μm the permeability value cannot be correctly calculated, as the rapid changes in impedance with frequency are consistent with a negative average permeability of the wire. This dependence can arise due to large heterogeneity of the permeability as well as the contribution of regions with axial anisotropy. As the magnitude of stresses affects the local anisotropy [32,33], the permeability is also sensitive to the stress distribution. In the case of large internal stresses, such as in thin wires [51], the radial dependence of permeability can be very sharp, and averaging provides large errors. In this case, the permeability spatial distribution and field dependences have to be taken into account in Maxwell's equations, and the first approximation is not applicable. For the samples with a diameter of 8.5 and 10 μm, the calculated permeabilities have relatively low values at zero DC magnetic field, and the radial dependence is not pronounced. The GMI curves measured for these samples correspond to the case of a radial magnetization distribution in the shell. The shapes of the hysteresis loops (which are presented in Figure 7 suggest, that radial anisotropy is formed in these samples, as the saturation field does not exceed 40 A/m. When the current passes through the sample, the processes of non-uniform magnetization rotation occurring in a radially magnetized shell may enhance losses and, thus, the imaginary part of the permeability. Since such processes are not taken into account in our model, the real values of permeability have to be much higher, than the calculated ones. It means, that such a simplified model cannot be used for microwires with a radially magnetized shell.

The permeability distribution calculated for the sample with a diameter of 8 μm shows a decreasing trend with increasing r. The sharp decrease in the permeability radial dependence occurs at a distance of around 0.60 of radius. The diameters of an axially magnetized core obtained from the squareness ratio and cross-magnetic moment were 0.67 a and 0.77 a, respectively. Consequently, an abrupt change in the circular permeability can be associated with a change of prevailing anisotropy type from axial to circular at the boundary between the core and the shell. For a more accurate determination of the core radius, fine sampling of the layers is required.

For a microwire with a diameter of 22 μm the circular permeability decreases monotonically with increasing r without sharp jumps. The absence of permeability jumps indicates, that the wire region under investigation have rather uniform micromagnetic structure. Presumably, all the averaging layers are located in the shell region. This result is also consistent with the core radii, obtained using the cross-magnetic moment or squareness ratio techniques.

For a sample with a metal diameter of 28 μm the permeability distribution has a sharp decrease at 0.82 of the wire radius. As the cross-magnetic moment signal indicated the absence of the core domain, and the squareness coefficient gave the value of ~0.14 a for the core domain radius, the results may be explained by the formation of the surface layer with much higher circular anisotropy.

A brief description of the results is presented in Table 3. Detailed table with the experimental results for all investigated samples is presented in the Supplementary Material.

Table 3. The results of indirect method of micromagnetic structure evaluation.

d_m [μm]	Structure	Method Applicability
6.4	Circular without core, sharp permeability radial dependence	Not applicable
8	Circular	Applicable, core radius calculated—0.60 R
8.5	Axial	Not applicable
10	Axial	Not applicable
22	Circular	Applicable, measurements in the shell
28	Circular	Applicable, non-uniform permeability

All the described techniques for estimating the core radius showed the consiste[n]t results in cases of the circular anisotropy in microwires. Nevertheless, there are seve[ral] subtleties in the applicability and accuracy of the methods, which have to be considere[d].

The impedance measurement in a wide frequency range with subsequent calculati[on] of the magnetic permeability distribution allow one to trace the changes of the magneti[za]tion distribution caused the presence of interfaces between regions with different type[s] anisotropy or defects. This method allows one to reconstruct the data on the microstructu[re] of a sample based on the experimental results, which makes it promising for microtomo[g]raphy of samples with a complex domain structure. The performed analysis shows, th[at] this model provides reasonable results in the case of circular anisotropy in the shell if t[he] current amplitude is low and does not cause the non-linear AC magnetization. In cas[es] of significant permeability variations with radial coordinate or radial/axial anisotro[py] additional magnetization mechanisms have to be taken into account.

4. Conclusions

In this paper we have presented a novel way to estimate the micromagnetic structu[re] in microwires, in particular, to determine the permeability distribution in the cross-secti[on] of the wire. The method is based on the dependence of the impedance of a cylindric[al] conductor on its circular permeability: from the frequency dependence of the impedan[ce] the permeability value for each current penetration depth was calculated in the first a[p]proximation of non-constant permeability effect on the impedance. Analysis of the radi[al] distribution of permeability made it possible to determine the position of boundary se[p]rating the core and shell of the wire with different anisotropy. The results were compar[ed] with those obtained by two magnetostatic methods involving different components of t[he] microwire magnetic moment. For this, the conventional hysteresis loops and the ma[g]netic field dependence of the cross-magnetic moment were measured. For the samp[le] with metallic core diameter of 8 μm and circular magnetization in the shell region t[he] calculated core radius corresponded to that obtained using two magnetostatic metho[ds]. Similar results were obtained for the thicker sample with 22 μm diameter. For the samp[le] with the largest diameter 28 μm the permeability have strongly non-uniform distributi[on] over the cross-section. For the rest of the samples the sharp permeability variations wi[th] radial coordinate and radial anisotropy limited the method applicability. For such sampl[e] simplified approach does not provide information about the permeability value, addition[al] magnetization mechanisms have to be taken into account.

It is concluded that the proposed method can be used for microwires with circul[ar] anisotropy and with a smooth change in permeability over the cross-section. The resul[ts] obtained correlate with the conventional method based on the squareness ratio estimati[on] but at the same time give additional information about the microstructure or local defec[ts] which has great potential for microtomography and defectoscopy.

Supplementary Materials: The following are available online at https://www.mdpi.com/2079[-] 991/11/2/274/s1, Table S1: Experimental data on the samples under investigation: magnetizati[on] curves, cross-magnetic moment dependences, GMI curves and the permeability radial dependen[ce].

Author Contributions: Conceptualization, N.P. and V.R. (Valeria Rodionova); methodology, N.[P.] validation, L.P., N.P.; formal analysis, I.A., N.P., L.P., V.R. (Valeria Rodionova); investigation, V.[R.] (Valeria Rodionova), V.R. (Vladimir Rodionov), V.K., I.A., N.A.; data curation, N.P., L.P., V.R. (Vale[ria] Rodionova); writing—original draft preparation, I.A.; writing—review and editing, V.R. (Vale[ria] Rodionova), L.P., N.P.; visualization, I.A.; supervision, N.P.; project administration, N.P.; fundi[ng] acquisition, N.P. All authors have read and agreed to the published version of the manuscript.

Funding: This research was funded by RFBR, grant number 18-02-00137. I.A. is supported by "BAS[IS"] foundation for the advancement of theoretical physics and mathematics. N.A., L.P. and I.A. we[re] supported from the Russian Academic Excellence Project at the Immanuel Kant Baltic Federal Universi[ty].

Informed Consent Statement: Not applicable.

Data Availability Statement: The data presented in this study are available in the article and Supplementary Material.

Conflicts of Interest: The authors declare no conflict of interest.

References

1. Coey, J.M.D. Amo rphous magnetic order. *J. Appl. Phys.* **1978**, *49*, 1646–1652. [CrossRef]
2. Alam, J.; Bran, C.; Chiriac, H.; Lupu, N.; Óvári, T.A.; Panina, L.V.; Rodionova, V.; Varga, R.; Vázquez, M.; Zhukov, A. Cylindrical micro and nanowires: Fabrication, properties and applications. *J. Magn. Magn. Mater.* **2020**, *513*, 167074. [CrossRef]
3. Zhukova, V.; Ipatov, M.; Zhukov, A. Thin Magnetically Soft Wires for Magnetic Microsensors. *Sensors* **2009**, *9*, 9216–9240. [CrossRef] [PubMed]
4. Zhukov, A.; Ipatov, M.; Churyukanova, M.; Talaat, A.; Blanco, J.M.; Zhukova, V. Trends in optimization of giant magnetoimpedance effect in amorphous and nanocrystalline materials. *J. Alloy. Compd.* **2017**, *727*, 887–901. [CrossRef]
5. Rodionova, V.; Ilyn, M.; Ipatov, M.; Zhukova, V.; Perov, N.; Zhukov, A. Spectral properties of electromotive force induced by periodic magnetization reversal of arrays of coupled magnetic glass-covered microwires. *J. Appl. Phys.* **2012**, *111*, 07E735. [CrossRef]
6. Hayashi, M.; Thomas, L.; Moriya, R.; Rettner, C.; Parkin, S.S.P. Current-Controlled Magnetic Domain-Wall Nanowire Shift Register. *Science* **2008**, *320*, 209–211. [CrossRef]
7. Kuzminski, M.; Nesteruk, K.; Lachowicz, H.K. Magnetic field meter based on giant magnetoimpedance effect. *Sens. Actuat. A Phys.* **2008**, *141*, 68–75. [CrossRef]
8. Li, D.R.; Lu, Z.C.; Zhou, S.X. Magnetic anisotropy and stress-impedance effect in Joule heated Fe73.5Cu1Nb3Si13.5B9 ribbons. *J. Appl. Phys.* **2004**, *95*, 204207. [CrossRef]
9. Zhukov, A. Domain wall propagation in a Fe-rich glass-coated amorphous microwire. *Appl. Phys. Lett.* **2001**, *78*, 3106–3108. [CrossRef]
10. Vázquez, M.; Chiriac, H.; Zhukov, A.; Panina, L.; Uchiyama, T. On the state-of-the-art in magnetic microwires and expected trends for scientific and technological studies. *Phys. Status Solidi A* **2010**, *207*, 1. [CrossRef]
11. Panina, L.V.; Mohri, K. Magnetoimpedance effect in amorphous wires. *Appl. Phys. Lett.* **1994**, *65*, 11891191. [CrossRef]
12. Sartorelli, M.L.; Knobel, M.; Schoenmaker, J.; Gutierrez, J.; Barandiaran, J.M. Giant magneto-impedance and its relaxation in CoFeSiB amorphous ribbons. *Appl. Phys. Lett.* **1997**, *71*, 22082210. [CrossRef]
13. Zhukov, A.; González, J.; Vázquez, M.; Larin, V.; Torcunov, A. Nanocrystalline and Amorphous Magnetic Microwires. In *Encyclopedia of Nanoscience and Nanotechnology*; Nalwa, H.S., Ed.; American Scientific: Stevenson Ranch, CA, USA, 2014; Volume 6, pp. 365–367.
14. Rodionova, V.; Baraban, I.; Chichay, K.; Litvinova, A.; Perov, N. The stress components effect on the Fe-based microwires magnetostatic and magnetostrictive properties. *J. Magn. Magn. Mater.* **2017**, *422*, 216–220. [CrossRef]
15. Zhukova, V.; Blanco, J.M.; Corte-Leon, P.; Ipatov, M.; Churyukanova, M.; Taskaev, S.; Zhukov, A. Grading the magnetic anisotropy and engineering the domain wall dynamics in Fe-rich microwires by stress-annealing. *Acta Mater.* **2018**, *155*, 279–285. [CrossRef]
16. Zhukova, V.; Umnov, P.; Molokanov, V.; Shalygin, A.N.; Zhukov, A.P. Studies of Magnetic Properties of Amorphous Microwires Produced by Combination of by Quenching, Glass Removal and Drawing Techniques. *Key Eng. Mater.* **2011**, *495*, 280–284. [CrossRef]
17. Baraban, I.; Panina, L.; Litvinova, A.; Rodionova, V. Effect of glass-removal on the magnetostriction and magnetic switching properties in amorphous FeSiB microwires. *J. Magn. Magn. Mater.* **2019**, *481*, 50–54. [CrossRef]
18. Panina, L.V.; Dzhumazoda, A.; Evstigneeva, S.A.; Adam, A.M.; Morchenko, A.T.; Yudanov, N.A.; Kostishyn, V.G. Temperature effects on magnetization processes and magnetoimpedance in low magnetostrictive amorphous microwires. *J. Magn. Magn. Mater.* **2018**, *459*, 147–153. [CrossRef]
19. Chiriac, H.; Óvári, T.A.; Pop, G.; Barariu, F. Magnetic behavior of nanostructured glass covered metallic wires. *J. Appl. Phys.* **1997**, *81*, 5817. [CrossRef]
20. Azuma, D.; Hasegawa, R.; Saito, S.; Takahashi, M. Effect of residual strain in Fe-based amorphous alloys on field induced magnetic anisotropy and domain structure. *J. Appl. Phys.* **2013**, *113*, 17A339. [CrossRef]
21. Hasegawa, R.; Takahashi, K.; Francoeur, B.; Couture, P. Magnetization kinetics in tension and field annealed Fe-based amorphous alloys. *J. Appl. Phys.* **2013**, *113*, 17A312. [CrossRef]
22. Usov, N.A.; Antonov, A.S.; Lagar'kov, A.N.; Granovsky, A.B. GMI spectra of amorphous wires with different types of magnetic anisotropy in the core and the shell regions. *J. Magn. Magn. Mater.* **1999**, *203*, 108–110. [CrossRef]
23. Donnely, C.; Guizar-Sicairos, M.; Scagnoli, V.; Gliga, S.; Holler, M.; Raabe, J.; Heyderman, L.J. Three-dimensional magnetization structures revealed with X-ray vector nanotomography. *Nature* **2017**, *547*, 328–331. [CrossRef]
24. Chizhik, A.; Zhukov, A.; González, J.; Gawroński, P.; Kułakowski, K. Spiral magnetic domain structure in cylindrically-shaped microwires. *Sci. Rep.* **2018**, *8*, 15090. [CrossRef]
25. González, J.; Chizhik, A.; Zhukov, A.; Blanco, J.M. Surface magnetization reversal and magnetic domain structure in amorphous microwires. *Phys. Status Solidi A* **2011**, *208*, 502–508. [CrossRef]
26. Zhukov, A.; Blanco, J.M.; Chizik, A.; Ipatov, M.; Rodionova, V.; Zhukova, V. Manipulation of domain wall dynamics in amorphous microwires through domain wall collision. *J. Appl. Phys.* **2013**, *114*, 043910. [CrossRef]
27. Zhukov, A.P.; Vázquez, M.; Velazquez, J.; Chiriac, H.; Larin, V. The remagnetization process in thin and ultra-thin Fe-rich amorphous wires. *J. Magn. Magn. Mater.* **1995**, *151*, 132–138. [CrossRef]

28. Rodionova, V.; Ipatov, M.; Ilyn, M.; Zhukova, V.; Perov, N.; González, J.; Zhukov, A. Tailoring of Magnetic Properties of Magnetostatically-Coupled Glass-Covered Magnetic Microwires. *J. Supercond. Nov. Magn.* **2011**, *24*, 541–547. [CrossRef]
29. Makhnovskiy, D.P.; Panina, L.V.; Mapps, D.J. Field-dependent surface impedance tensor in amorphous wires with two types of magnetic anisotropy: Helical and circumferential. *Phys. Rev. B* **2001**, *63*, 144424. [CrossRef]
30. Zhukov, A.; Zhukova, V.; Blanco, J.M.; Cobeño, A.F.; Vázquez, M.; González, J. Magnetostriction in glass-coated magnetic microwires. *J. Magn. Magn. Mater.* **2003**, *258*, 151–157. [CrossRef]
31. Ekstrom, P.A.; Zhukov, A. Spatial structure of the head-to-head propagating domain wall in glass-covered FeSiB microwire. *Phys. D Appl. Phys.* **2010**, *43*, 205001. [CrossRef]
32. Chiriac, H.; Óvári, T.A.; Pop, G. Internal stress distribution in glass-covered amorphous magnetic wires. *Phys. Rev. B* **1995**, *5*, 10104–10113. [CrossRef] [PubMed]
33. Antonov, A.S.; Borisov, V.T.; Borisov, O.V.; Prokoshin, A.F.; Usov, N.A. Residual quenching stresses in glass-coated amorphous ferromagnetic microwires. *J. Phys. D Appl. Phys.* **2000**, *33*, 1161–1168. [CrossRef]
34. Chizhik, A.; Garcia, C.; Zhukov, A.; Gawroński, P.; Kułakowski, K.; González, J.; Blanco, J.M. Investigation of helical magnetic structure in Co-rich amorphous microwires. *J. Magn. Magn. Mater.* **2007**, *316*, 332–336. [CrossRef]
35. Chizhik, A.; Blanco, J.M.; Zhukov, A.; González, J.; Garcia, C.; Gawroński, P.; Kulakowski, K. Magneto-optical determination of helical magnetic structure in amorphous microwires. *Phys. B Condens. Matter* **2008**, *403*, 289–292. [CrossRef]
36. Severino, A.M.; Gómez-Polo, C.; Marín, P.; Vázquez, M. Influence of the sample length on the switching process of magnetostrictive amorphous wire. *J. Magn. Magn. Mater.* **1992**, *103*, 117–125. [CrossRef]
37. Liu, K.; Lu, Z.; Liu, T.; Li, D. Measurement of internal tensile stress in $Co_{68.2}Fe_{4.3}Cr_{3.5}Si_{13}B_{11}$ glass-coated amorphous microwire using the stress sensitivity of saturation magnetostriction. *J. Magn. Magn. Mater.* **2013**, *339*, 124–126. [CrossRef]
38. Panina, L.; Dzhumazoda, A.; Nematov, M.; Alam, J.; Trukhanov, A.; Yudanov, N.; Morchenko, A.; Rodionova, V.; Zhukov, A. Soft Magnetic Amorphous Microwires for Stress and Temperature Sensory Applications. *Sensors* **2019**, *19*, 5089. [CrossRef] [PubMed]
39. Vázquez, M. Advanced Magnetic Microwires. In *Handbook of Magnetism and Advanced Magnetic Materials*; Kronmüller, H., Parkin, S., Eds.; Wiley: Chichester, UK, 2007; Volume 4, pp. 2193–2296.
40. Larin, V.S.; Torcunov, A.V.; Zhukov, A.; González, J.; Vázquez, M.; Panina, L. Preparation and properties of glass-coated microwires. *J. Magn. Magn. Mater.* **2002**, *249*, 39–45. [CrossRef]
41. Rodionova, V.V.; Baraban, I.A.; Panina, L.V.; Bazlov, A.I.; Perov, N.S. Tunable magnetic properties of glass-coated microwires by initial technical parameters. *IEEE Trans. Magn.* **2018**, *54*, 2002706. [CrossRef]
42. Buznikov, N.A.; Antonov, A.S.; Granovsky, A.B. Asymmetric magnetoimpedance in amorphous microwires due to bias current: Effect of torsional stress. *J. Magn. Magn. Mater.* **2014**, *355*, 289–294. [CrossRef]
43. Chizhik, A.; Garcia, C.; Zhukov, A.; González, J.; Gawroński, P.; Kułakowski, K.; Blanco, J.M. Relation between surface magnetization reversal and magnetoimpedance in Co-rich amorphous microwires. *J. Appl. Phys.* **2008**, *103*, 1–4. [CrossRef]
44. Sossmeier, K.D.; Oliveira, J.T.D.; Schelp, L.F.; Carara, M. Domain wall dynamics in CoFeSiB microwires under axial applied stress. *J. Magn. Magn. Mater.* **2007**, *316*, e541–e544. [CrossRef]
45. Dionne, G.F. Magnetic relaxation and anisotropy effects on high-frequency permeability. *IEEE Trans. Magn.* **2003**, *3*, 3121–3126. [CrossRef]
46. Astefanoaei, I.; Radu, D.; Chiriac, H. Internal stress distribution in DC joule-heated amorphous glass-covered microwires. *J. Phys. Cond. Matter.* **2006**, *18*, 2689–2716. [CrossRef]
47. Torrejon, J.; Thiaville, A.; Adenot-Engelvin, A.L.; Vázquez, M.; Acher, O. Cylindrical magnetization model for glass-coated microwires with circular anisotropy: Statics. *J. Magn. Magn. Mater.* **2011**, *323*, 283–289. [CrossRef]
48. Perov, N.; Radkovskaya, A.A. Vibrating sample anisometer. In *Proceeding of 1&2 Dimensional Magnetic Measurements and Testing*; Vienna Magnetic Group Report: Vienna, Austria, 2001; pp. 104–108.
49. Vázquez, M.; Zhukov, A.P.; Garcia, K.L.; Pirota, K.R.; Ruiz, A.; Martinez, J.L.; Knobel, M. Temperature dependence of magnetization reversal in magnetostrictive glass-coated amorphous microwires. *Mat. Sci. Eng. A* **2004**, *375–377*, 1145–1148.
50. Barandiaran, J.M.; Hernando, A.; Madurga, V.; Nielsen, O.V.; Vázquez, M.; Vázquez-Lopez, M. Temperature, stress, and structural-relaxation dependence of the magnetostriction in (Co0.94/BFe0.06)75/BSi15B10 glasses. *Phys. Rev. B* **1987**, *35*, 506. [CrossRef] [PubMed]
51. Churyukanova, M.; Semenkova, V.; Kaloshkin, S.; Shuvaeva, E.; Gudoshnikov, S.; Zhukova, V.; Shchetinin, I.; Zhukov, A. Magnetostriction investigation of soft magnetic microwires. *Phys. Status Solidi A* **2016**, *213*, 363–367. [CrossRef]
52. Chichay, K.; Rodionova, V.; Zhukova, V.; Kaloshkin, S.; Churyuknova, M.; Zhukov, A. Investigation of the magnetostriction coefficient of amorphous ferromagnetic glass coated microwires. *J. Appl. Phys.* **2014**, *116*, 173904. [CrossRef]
53. Nematov, M.G.; Baraban, I.; Yudanov, N.A.; Rodionova, V.; Qin, F.X.; Peng, H.-X.; Panina, L.V. Evolution of the magnetic anisotropy and magnetostriction in Co-based amorphous alloys microwires due to current annealing and stress-sensory applications. *Alloy. Comp.* **2020**, *837*, 155584. [CrossRef]
54. Liu, K.; Lu, Z.; Liu, T.; Li, D. Influence of Tensile Force During Preparation on Internal Stress of Glass-Coated Microwires. *Supercond. Nov. Magn.* **2013**, *26*, 2969–2973. [CrossRef]
55. Zhukov, A.; Chichay, K.; Talaat, A.; Rodionova, V.; Blanco, J.M.; Ipatov, M.; Zhukova, V. Manipulation of magnetic properties of glass-coated microwires by annealing. *J. Magn. Magn. Mater.* **2015**, *383*, 232–236. [CrossRef]
56. Vázquez, M. Giant magneto-impedance in soft magnetic "Wires". *J. Magn. Magn. Mater.* **2001**, *226–230*, 693–699.

 *nanomaterials*

ticle

High-Frequency Magnetoimpedance (MI) and Stress-MI in Amorphous Microwires with Different Anisotropies

naid Alam [1], Makhsudsho Nematov [2,3], Nikolay Yudanov [1], Svetlana Podgornaya [1] and Larissa Panina [1,2,*]

1. Institute of Novel Materials and Nanotechnology, National University of Science and Technology, MISiS, 119991 Moscow, Russia; engr.sajalam@gmail.com (J.A.); kolyan2606@mail.ru (N.Y.); podgsv@mail.ru (S.P.)
2. Institute of Physics, Mathematics & IT, Immanuel Kant Baltic Federal University, 236041 Kaliningrad, Russia; nmg2409@gmail.com
3. Faculty of Energy, Tajik Technical University Named after ac. M.S. Osimi, Dushanbe 734042, Tajikistan
* Correspondence: drlpanina@gmail.com; Tel.: +7-926-076-5513

Abstract: Magnetoimpedance (MI) in Co-based microwires with an amorphous and partially crystalline state was investigated at elevated frequencies (up to several GHz), with particular attention paid to the influence of tensile stress on the MI behavior, which is called stress-MI. Two mechanisms of MI sensitivity related to the DC magnetization re-orientation and AC permeability dispersion were discussed. Remarkable sensitivity of impedance changes with respect to applied tensile stress at GHz frequencies was obtained in partially crystalline wires subjected to current annealing. Increasing the annealing current enhanced the axial easy anisotropy of a magnetoelastic origin, which made it possible to increase the frequency of large stress-MI: for 90mA-annealed wire, the impedance at 2 GHz increased by about 300% when a stress of 450 MPa was applied. Potential applications included sensing elements in stretchable substrates for flexible electronics, wireless sensors, and tunable smart materials. For reliable microwave measurements, an improved SOLT (short-open-load-thru) calibration technique was developed that required specially designed strip cells as wire holders. The method made it possible to precisely measure the impedance characteristics of individual wires, which can be further employed to characterize the microwave scattering at wire inclusions used as composites fillers.

Keywords: amorphous microwires; high-frequency magnetoimpedance; SOLT calibration; magnetic anisotropy

ation: Alam, J.; Nematov, M.; danov, N.; Podgornaya, S.; Panina, High-Frequency Magnetoimpedance) and Stress-MI in Amorphous crowires with Different isotropies. *Nanomaterials* **2021**, *11*, 8. https://doi.org/10.3390/ no11051208

ademic Editor: Matt Cole

ceived: 2 March 2021
cepted: 23 April 2021
blished: 2 May 2021

blisher's Note: MDPI stays neutral h regard to jurisdictional claims in blished maps and institutional affil- ons.

1. Introduction

Soft ferromagnetic microwires have attracted growing attention in research since they exhibit a number of physical effects important for applications (see, for example, recent reviews [1–3]). Controlling their microstructure, chemical composition, and phase composition, cross-sectional size allows for the development of novel concepts for their applications in sensory devices and functional materials. Here, we investigated high frequency magnetoimpedance (MI) in Co-based microwires with amorphous and partially crystalline states, paying particular attention to the influence of a tensile stress on the MI behavior at elevated frequencies. Potential applications include sensing elements in stretchable substrates for flexible electronics, wireless sensors, and tunable smart materials [4–6].

The MI effect, which is referred to as a large and sensitive change in the complex-valued impedance of a ferromagnetic conductor in the presence of external magnetic field and other stimuli such as a mechanical stress and temperature, led to the development of various sensors driven by alternative current (AC) or pulse circuits operating at megahertz frequencies [7–11]. A high sensitivity of MI with respect to the external factors is caused by the directional change in the quasi-static (DC) magnetization. Therefore, it was expected that MI remains sensitive for higher frequencies in the GHz range. At high frequencies,

when the skin effect is strong, the impedance dependence on the magnetic parameters h
the form [12]:

$$Z = Z_c \left(\sqrt{\mu_{ef}} \cos^2 \theta + \sin^2 \theta \right)$$

Here, Z_c is the impedance of a non-ferromagnetic conductor, μ_{ef} is the AC permeabili
parameter, and θ is the angle between the static magnetization and high-frequency curre
Equation (1) shows that as long as the angle θ varies in response to external factors ar
the value of μ_{ef} differs from unity, the high frequency impedance demonstrates $\cos^2$
dependence. The permeability also depends on θ and other DC parameters, but th
dependence is typically weak since high values of a DC magnetic field and/or anisotro
are required to satisfy the condition of the ferromagnetic resonance at these frequencies [1
A typical magnetic field behavior of impedance at GHz frequencies in Co-rich amorpho
microwires with a well-formed circumferential easy anisotropy is given in Figure 1. T
inset shows the DC magnetization loop, and thus it is clearly seen that the field interv
where the magnetization rotates towards the axis corresponded to a sharp increase
impedance. When the static magnetization was along the wire ($\theta = 0$) and did not chan
by the field, only the permeability parameter $\mu_{ef}(H)$ determined the field behavior of t
impedance: it increased with much lower slope, and an inflexion point was seen in the pl
of $Z(H)$. In this field range, the sensitivity of the impedance change was low, of about
few percent per Oe. Similar high-frequency impedance behaviors in these types of wir
were demonstrated in a number of publications [14–16].

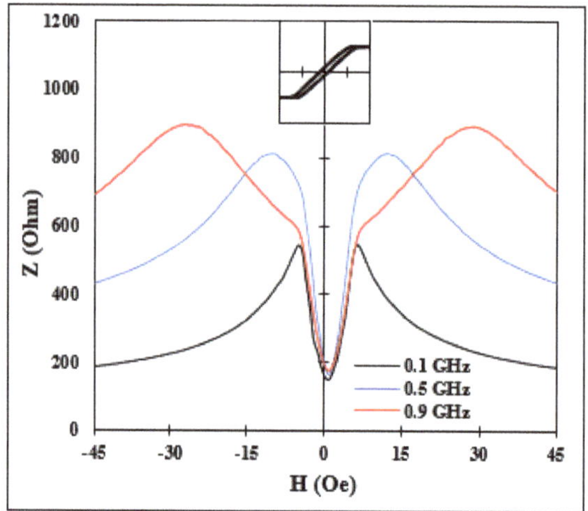

Figure 1. Typical MI in wires with circumferential anisotropy. For this measurement, we us
amorphous $Co_{66.6}Fe_{4.28}B_{11.51}Si_{14.48}Ni_{1.44}Mo_{1.69}$ microwires. Inset shows the DC loop along the wi

To enhance the MI sensitivity, researchers have proposed various annealing regimes
order to establish a required transverse anisotropy of small amplitude and small easy-ax
deviations [17–21]. In the case of stress-MI, the induced circumferential anisotropy
combination with positive magnetostriction formed by current annealing leads to a lar
change in impedance in response to tensile stress without use of a bias magnetic field [21,2
This effect is of high practical importance; however, the impedance vs. stress sensitivi
can quickly decrease with increasing frequency. Sensitive change in the DC magnetizatio
requires a small anisotropy; then, the ferromagnetic frequency is low, and the frequen
range of hundred MHz corresponds to the tail of the ferromagnetic resonance. As follov
from Equation (1), if the values of μ_{ef} are close to unity, the DC magnetic configuratio
does not affect the impedance.

An interesting situation can occur in wires with a non-uniform-induced anisotropy or with a relatively high axial anisotropy. Thus, large variations of impedance under the effect of mechanical stress can be expected if it causes the anisotropy regions redistribution. In the case of high axial anisotropy of magnetoelastic origin, the ferromagnetic resonance frequency is in the GHz range, with the external stress changes this frequency causing large changes in stress-MI. Here, we investigated the effect of tensile stress on high-frequency impedance in wires with different anisotropy and magnetostriction modified by current annealing. It was found that modifications in the axial anisotropy caused by applied stress led to a large impedance change at high frequencies due to shift in the resonance frequency.

To optimize the impedance behavior, one must measure the impedance characteristics of individual wires that may present considerable problems at higher frequencies in the GHz range. This is related to uncertainties occurring due to calibration of measuring devices (network analyzer) with customized sample arrangements. In this case, customized calibration methods should be used. Coaxial and microstrip transmission lines are widely used as sample holders [23,24]; however, in the case of the impedance measurements when the sample is subjected to mechanical stress, coaxial lines are not convenient. Here, we discuss the calibration procedures for microstrip lines. In the microwave de-embedding methods, the electrical reference planes are shifted to the sample location. This requires the use of dummy sample holders with specific terminations, for example, three reflection standards (open, short, and match load) and two ports connected together (thru). This calibration procedure is known by the name of short-open-load-thru (SOLT) or thru-open-short-match (TOSM) [25]. Accurate calibration standards can be fabricated on the customized printed circuit board (PCB), and then the complex impedance Z is calculated from S_{21} parameter using a simple formula without solving any equations.

The accuracy of the customized calibration methods critically depends on the tolerance of calibration standards fabrication (for example, the lumped open, short, etc., terminations) and the problem becomes more essential with increasing frequency. Relating to this, an alternative of VNA calibration technique for MI measurements at a high frequency was recently introduced in [26] using zero-field de-embedding method. In this method, the MI measurements obtained at zero magnetic field were used as a reference signal, and then the same process was repeated with applied magnetic field so that the data could be subtracted by the values of the reference signal. However, the major disadvantage of this method is the absence of zero-field behavior. In this work, the SOLT calibration standard was adjusted by using specially designed strip cells as a PCB SOLT standard for VNA calibration and measuring the high frequency MI and stress-MI, which could be directly used for the development of various high-performance micro-sensor devices.

2. Materials and Methods

2.1. Wire Samples

Glass-coated amorphous microwires of two Co-based compositions $Co_{66.6}Fe_{4.28}B_{11.51}Si_{14.48}Ni_{1.44}Mo_{1.69}$ and $Co_{71}Fe_5B_{11}Si_{10}Cr_3$ produced by modified Taylor–Ulitovskiy method [27,28] were used in this work. The wires of the first composition had a total diameter $D = 25.8$ μm and a metal core diameter $d = 14.2$ μm (referred to as sample No1). The wire samples of the second composition had different geometry: $D = 30$ μm, $d = 25$ μm and $D = 43$ μm, $d = 35.2$ μm (referred to as samples No2 and No3, respectively).

The wires with smaller diameter were fully amorphous, whereas the wires with larger diameter (sample No 3) were partially crystallized, as was concluded from DSC (differential scanning calorimetry) analysis in previous works [29,30]. Co-based wires typically have a negative magnetostriction and a circular type of the easy anisotropy since the internal stress is predominantly axial. Samples No 1 with a larger ratio of Co/Fe~15 content had the magnetostriction constant of about $-1.2 \cdot 10^{-7}$ [30,31] and a circumferential easy anisotropy. Sample No2 had a smaller negative magnetostriction ($<10^{-7}$ [21,32]) but an axial easy anisotropy. Sample No3 had a large and positive magnetostriction constant of about 10^{-6} due to partial crystallization [32] and an axial easy anisotropy. The crystallization and

Curie temperatures of the studied samples determined from the DSC curves using standar
software application were about 790 and 720–730 K, respectively.

The wire samples were annealed by DC current in air atmosphere with the aim t
investigate the effect of anisotropy and magnetostriction change on the MI and stress-M
behavior. The sample length for annealing was 15 cm, the DC current intensity was chose
between 25–90 mA, and the flowing current time was 30 min. All current treatments wei
performed in the same ambient conditions. The current magnitude was chosen to realize
heating effect in the range of temperatures 450–750 K. For moderate annealing curren
the annealing temperature was measured using the temperature dependence of resistivi
in a bridge circuit. For higher annealing currents (>60 mA), the annealing temperature wa
estimated from modelling on the basis of the energy balance. The details are given in [2]

2.2. DC Magnetic Measurements

The magnetic hysteresis loops were determined by an inductive method utilizin
two differential detection coils with 3 mm in the inner diameter and 5 cm in length. Th
magnetizing field in the direction of the wire length (axial direction) with a frequenc
of 500 Hz and amplitude of up to 1000 A/m was sufficient for re-magnetizing the so
magnetic wires. To increase the accuracy of the magnetization loop determination, we use
digital integration of the induced voltage.

2.3. Impedance Measurements

To obtain reliable measurement results at high frequencies by SOLT calibration, w
used specially designed strip cells, shown in Figure 2, as calibration standards. Th
calibration process is the same as for standard SOLT/TOSM [25,33], but instead of coaxi
standards, we used customized strip cells (PCB board). The microwire was installed in th
measuring cell of PCB. This calibration technique allows one to move the reference plane
towards the ends of the wire sample and only measure its S-parameters.

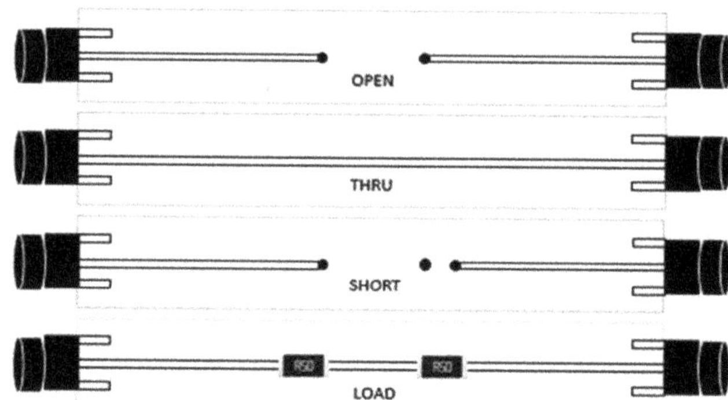

Figure 2. Schematic of specially designed PCB calibration cells with the SOLT standards. THR
standard utilizes direct connection ("flush THRU") with $S_{21} = S_{12} = 1$, $S_{11} = 0$, $S_{22} = 0$; SHORT
realized via connecting the signal strip to the ground; LOAD utilizes two 50 Ω RF Vishay resisto
connected to the ground in parallel.

For the impedance measurements, the standards were mounted on PCB as demoi
strated in Figure 3. The PCB calibration and measuring cells utilized FR4 material (glas
reinforced epoxy laminate) with a relative permittivity of 4.35 for frequencies highe
than 500 MHz. It was assumed that $S_{11,\,open} = 1$, $S_{11,\,short} = -1$, $S_{11,\,load} = 0$ fc
any frequencies. For THRU, a direct connection of the signal stripes was used, the
$S_{21,\,thru} = S_{12,\,thru} = 1$, $S_{11,\,thru} = 0$, $S_{22,\,thru} = 0$.

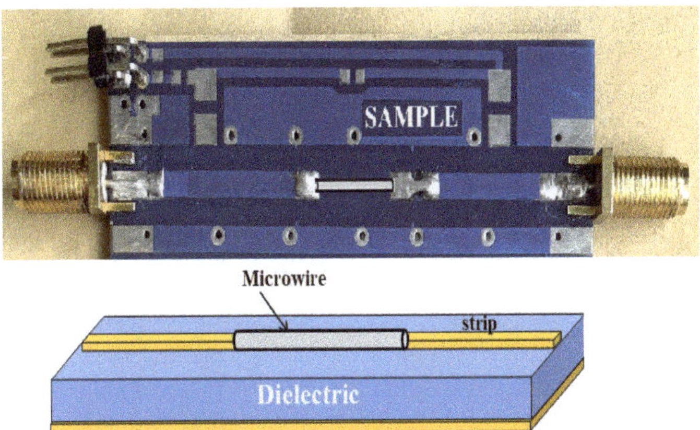

Figure 3. Wire sample as a planar waveguide resulting in a phase incursion: upper figure is a sample holder photo; bottom figure represents a schematic illustration.

In the microwave region, the length of electromagnetic waves along the sample may become of the order of its length, and thus the sample impedance would no longer be regarded as lumped, and hence the deduced impedance value will contain a significant error. The correction can be made considering that the sample placed between the stripes above the dielectric with the continuous ground on its opposite side represents a planar waveguide, as shown in Figure 3 [16]. The difference from the usual strip line construction is only due to the geometry of a conductor: a wire instead of a strip.

Therefore, this wire waveguide can be characterized by standard parameters: losses and delay time. The losses must be attributed to the intrinsic physical properties of the sample. On the contrary, the delay time Δt is caused mostly by the length of the sample. However, it introduces significant phase distortions at higher frequencies:

$$S_{21}(\omega) = \widetilde{S}_{21} \exp(j\gamma(\omega)) \exp(-j\omega\Delta t) \tag{2}$$

where $\widetilde{S}_{21}$ is the wire lamped S_{21} parameter, $\omega = 2\pi f$ is the angular frequency, j is the imaginary unit, $\gamma(\omega)$ is a phase function contributing to internal impedance properties, and $\exp(-i\omega\Delta t)$ is responsible for the phase incursion due to the time delay Δt. Thus, after measuring Δt along the sample, its intrinsic $\widetilde{S}_{21}$ parameter can be recovered from the measured one by multiplying by $\exp(j\omega\Delta t)$.

This parameter is already lumped from the point of view of the propagating wave, although it depends on the length of the sample. Using Equation (2), we can calculate the value of impedance, free of the waveguide properties, as

$$Z(\omega, H) = \frac{100 \cdot (1 - \widetilde{S}_{21}(\omega, H))}{\widetilde{S}_{21}(\omega, H)} \tag{3}$$

Figure 4 shows an example of the impedance dispersion of an amorphous microwire, measured with the modified SOLT calibration (user's ideal calibration kit created on VNA).

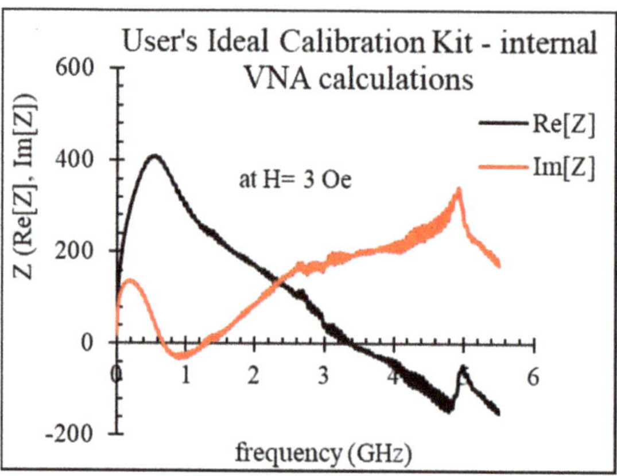

Figure 4. Dispersions of the impedance of amorphous $Co_{71}Fe_5B_{11}Si_{10}Cr_3$ microwire (sample N$\cdots$ as-cast) obtained from using user's ideal calibration kit created on VNA.

With increasing the frequency, the real part of the impedance starts to decrea$\cdots$ becoming negative, while the imaginary part grows up, becoming positive. This n$\cdots$ physical behavior is caused by the time delay. For removing this effect, we applied $\cdots$ additional adjustment described above. The impedance spectra after phase compensati$\cdots$ are shown in Figure 5. The resonance seen at a frequency of about 5 GHz is a parasi$\cdots$ electric resonance related to the OPEN terminations on the PCB calibration cell. It cann$\cdots$ be removed, and a sort of smooth interpolation is needed in this frequency range.

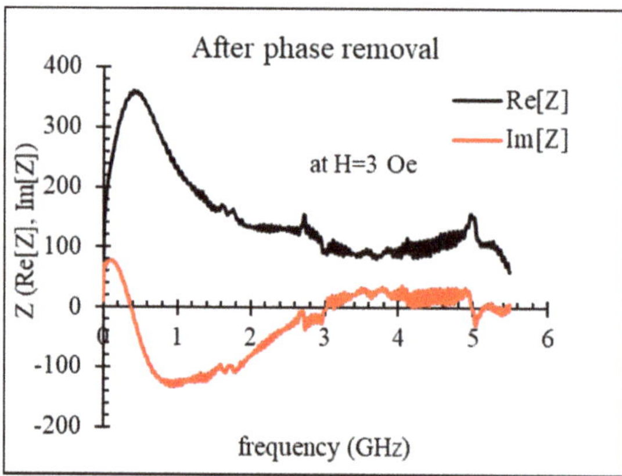

Figure 5. Dispersions of the impedance of amorphous $Co_{71}Fe_5B_{11}Si_{10}Cr_3$ microwire (sample N$\cdots$ as-cast) obtained from using user's ideal calibration kit created on VNA after phase compensatio$\cdots$

The parasitic resonances observed between 5 and 6 GHz were introduced during t$\cdots$ SOLT calibration and were caused by the PCB cell and connection lines. These resonanc$\cdots$ are well localized and can be easily replaced with smooth interpolants.

The Hewlett-Packard 8753E Vector Network Analyzer was used to measure the wi$\cdots$ impedance in the frequency range between 0.1 and 2.5 GHz at room temperature. The wi$\cdots$ length for impedance measurements was 11 mm. The impedance spectra were deduc$\cdots$ from S_{21} parameter (two-port measuring scheme) using Equation (3). The sample w$\cdots$ placed inside a Helmholtz coil that produced a slowly varying magnetic field up to 50 C$\cdots$ In order to introduce a tensile stress during S_{21} measurement, we hanged the load wi$\cdots$

a thread in the middle part of the wire. The load weight was applied to the whole wire, and thus the stress in the metallic core should be evaluated considering different Young's moduli of the metal and glass [34].

3. Results and Discussions

3.1. DC Properties

We investigated the wires of two Co-based compositions with very similar relative content of Co and Fe: 15.6 and 14.2, respectively. It is known that if this ratio is larger than 14, the magnetostriction is negative [31]. Therefore, it could be expected that all the samples have a negative magnetostriction, a circumferential easy anisotropy, and inclined hysteresis loops. However, only sample No1 showed an expected hysteresis loop in as-cast state as shown in Figure 6.

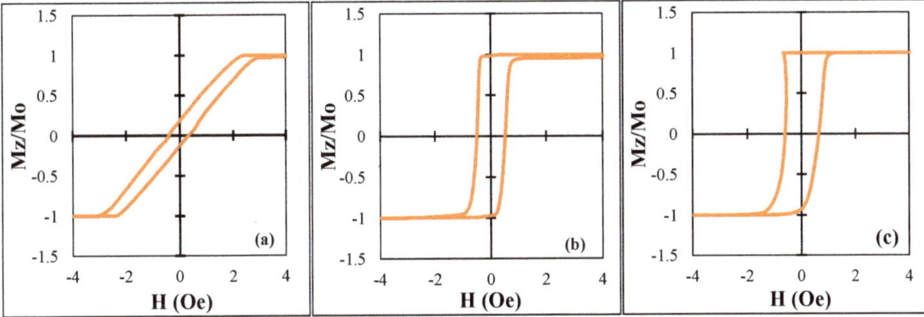

Figure 6. Hysteresis loops in a magnetic field along the wire axis for all the samples: (**a**) −1, (**b**) −2, and (**c**) −3. M_z is the magnetization component along the wire, M_0 is the saturation magnetization.

Samples No2 and No3 had rectangular hysteresis loops, which is associated with an axial easy anisotropy. In the case of sample No2, the magnetostriction constant was negative but very small, in the order of -10^{-7} [21]. The axial anisotropy could have been due to the existence of residual Co crystals with large magnetic anisotropy. In the case of sample No3, its magnetostriction was relatively large and positive (approximately $\approx 10^{-6}$ [32]) due to partial crystallization, and thus the loop had a rectangular shape. The loop appeared to be asymmetrical, but this was related to digital integration of a sharp voltage pulse induced during re-magnetization. In all the cases, we used current annealing to modify the anisotropy and magnetostriction. It is known that current annealing of amorphous wires when heating below the Curie temperature induces a circumferential easy anisotropy [35]. This was observed in wires with small magnetostriction, but in the case of high and positive magnetostriction, the induced circumferential anisotropy may be insufficient to overcome the magnetoelastic anisotropy and the easy axis of anisotropy would remain axial. In all the cases, higher annealing currents producing the heating above the Curie temperature did not form an induced circumferential anisotropy, but heating increased the magnetostriction constant. This strengthened the axial easy anisotropy. The results of the change in the hysteresis loops due to annealing with different current I_{an} are shown in Figure 7.

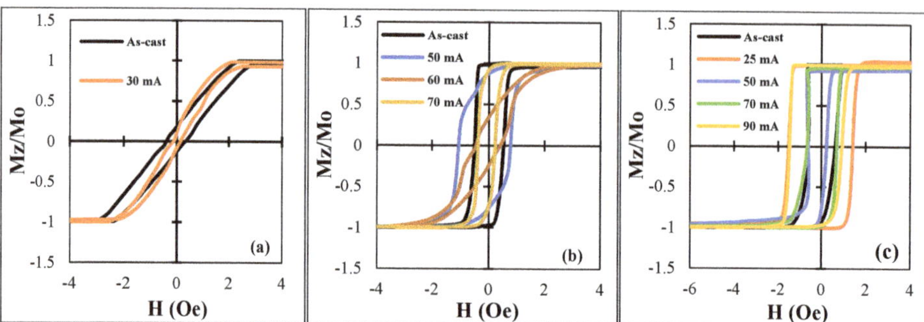

Figure 7. Hysteresis loops for all the samples after DC current annealing with different I_{an} for 30 min: (**a**) −1, (**b**) −2, and (**c**) −3.

3.2. Impedance vs. Field at High Frequencies

With this choice of wire samples, we had different combinations of easy anisotropy and magnetostriction. First, we examined the impedance frequency behavior. For a cast sample No1, the frequency plots are shown in Figure 1 and in Figure 8a for higher frequencies. In this case, large change in impedance due to magnetization re-orientation was observed for frequencies higher than 2 GHz, and the corresponding sensitivities were 42%/Oe at 1.5 GHz and 26.6%/Oe at 2.1 GHz.

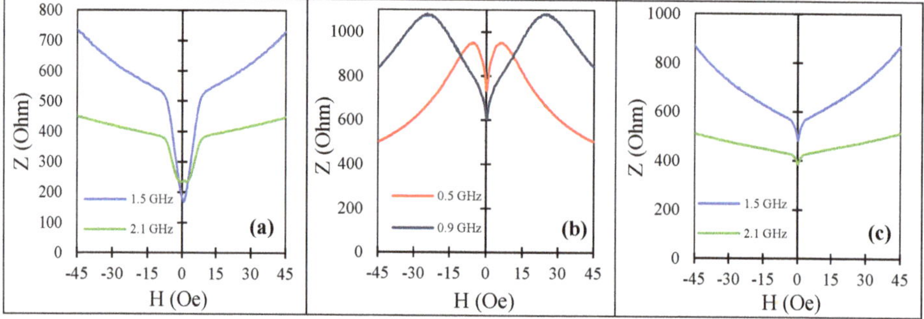

Figure 8. MI vs. magnetic field for amorphous $Co_{66.6}Fe_{4.28}B_{11.51}Si_{14.48}Ni_{1.44}Mo_{1.69}$ microwires (sample No1): (**a**) as-cast, (**b**,**c**) after current annealing with $I_{an} = 30$ mA for 30 min for different frequencies. Z corresponds to the impedance amplitude.

In this case, current annealing led to a combination of induced circumferential anisotropy and positive magnetostriction, which formed a helical type of easy anisotropy. The change in magnetostriction after annealing was due to structural relaxation and modification in atomic co-ordination [21,36]. The variation of impedance at low fields due to magnetization re-orientation was smaller and quickly decreased with increasing frequency, as seen in Figure 8b,c.

For a wire with an axial anisotropy (Figure 9, sample No2) the impedance vs. field dependence was only due to the permeability dispersion (see Equation (1)), and there was no sensitive impedance change at low frequencies where the MI plots had a single peak at zero field. With increasing frequency, two symmetrical peaks appeared and the field at maximum increased with increasing a frequency due to increase in the frequency of the ferromagnetic resonance. Nevertheless, at a frequency of about 1 GHz, the relative change in impedance ($\Delta Z / Z_0 = |(Z(H) - Z_0)/Z_0|$) $Z_0 = Z(H = 0)$ was about 10%/Oe. It was also noticed that in the case of an axial anisotropy, the impedance field sensitivity was higher for higher frequencies and remained at a level of 5–7%/Oe up to 1.5 GHz whilst the maximum sensitivity at 100 MHz was only 4.5%/Oe. Similar impedance characteristic were observed for sample No3, which also had an axial easy anisotropy. This behavior suggests that in the case of an axial anisotropy, the change in the anisotropy value may strongly influence the impedance at elevated frequencies.

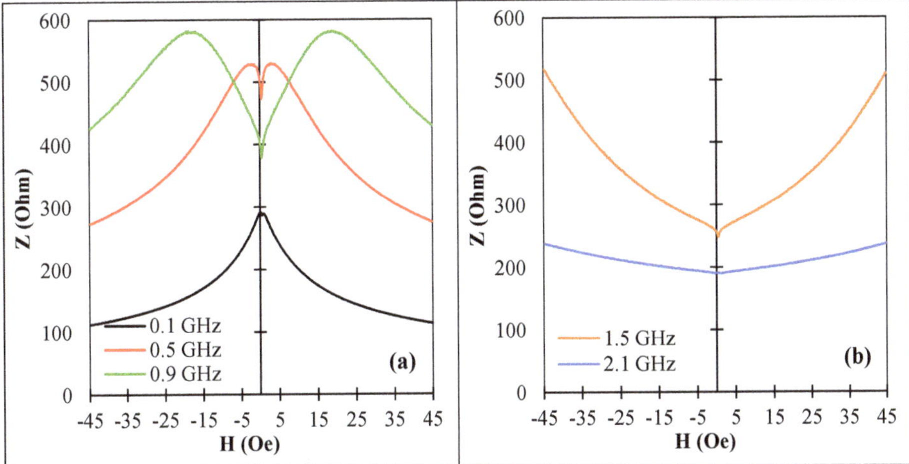

Figure 9. MI vs. magnetic field for amorphous $Co_{71}Fe_5B_{11}Si_{10}Cr_3$ microwires (sample No2) in as-prepared state for different frequencies: (**a**) 0.1 GHz to 0.9 GHz and (**b**) 1.5 GHz to 2.1 GHz. Z corresponds to the impedance amplitude.

After optimal current annealing, sample No2 had an induced circumferential anisotropy and positive magnetostriction. The impedance spectra are shown in Figure 10. The sensitive low-field region appeared and the maximal sensitivity at 1 GHz was about 50%/Oe and about 15%/Oe for a frequency of 1.5 GHz. These values were smaller than for sample No1 in as-prepared state since the induced anisotropy was of a helical type and its value was slightly lower. However, the combination of circumferential (or helical anisotropy) and positive magnetostriction was of interest to realize large impedance change at zero field [20,21], as is discussed in the next section.

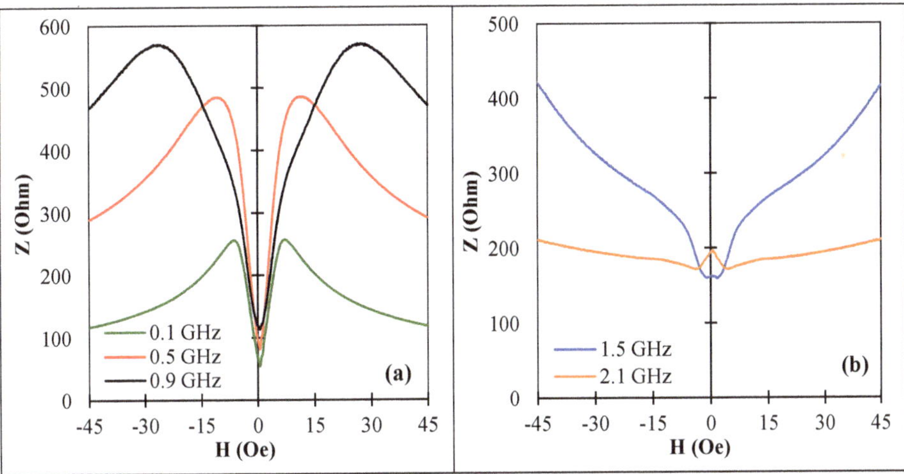

Figure 10. MI vs. magnetic field for amorphous $Co_{71}Fe_5B_{11}Si_{10}Cr_3$ microwires (sample No2) after current annealing with $I_{an} = 60$ mA for 30 min at GHz frequency range: (**a**) 0.1 GHz to 0.9 GHz and (**b**) 1.5 GHz to 2.1 GHz. Z corresponds to the impedance amplitude.

3.3. Effect of Tensile Stress on Impedance

In the case of a sample with a circumferential anisotropy and negative magnetostriction, the application of a tensile stress resulted in increase in the circumferential anisotropy and thus the main effect widened the impedance plots, as shown in Figure 11.

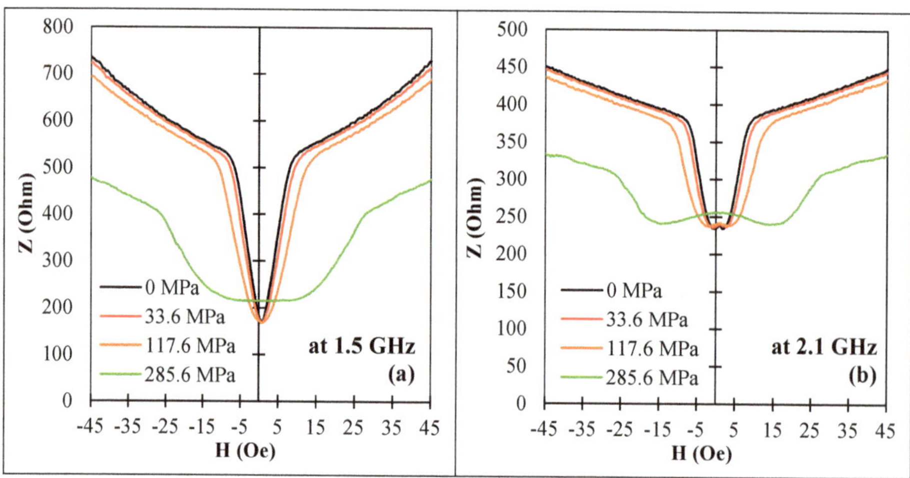

Figure 11. Effect of a tensile stress on the MI vs. magnetic field plots for amorphous $Co_{66.6}Fe_{4.28}B_{11.51}Si_{14.48}Ni_{1.44}Mo_{1.69}$ microwires (sample No1) in as-cast state at frequencies: (**a**) 1.5 GHz and (**b**) 2.1 GHz. Z corresponds to the impedance amplitude.

In this case, large impedance changes in wires subjected to tensile stress requires the application of a DC bias field.

For amorphous wires with the combination of axial anisotropy and negative magnetostriction (sample No2, as-prepared) and circumferential anisotropy and positive magnetostriction (sample No2, current annealed), the tensile stress affected the impedance even at zero magnetic field due to a change in the magnetization direction by stress, as shown Figure 12.

The strongest change in impedance caused by the application of tensile stress was observed for current annealed wire, which had a combination of induced circular anisotropy and positive magnetostriction. A large stress-MI effect preserved up to 1 GHz, as shown in Figure 13. For a range of 300 MPa, the impedance increased almost four times at frequency of 0.5 GHz, and it still increased by 60% at a frequency of 1.5 GHz.

In the case of an axial anisotropy and positive magnetostriction, the effect of tensile stress on MI was related to the anisotropy change and its influence on the permeability frequency dispersion. Unexpectedly, this could be even stronger than the effect magnetization re-orientation. At low frequencies, the impedance values decreased with increasing an axial anisotropy (that is, with increasing a tensile stress) since it was mainly determined by the behavior of the initial rotational permeability, which was inversely proportional to the anisotropy field. With increasing frequency, higher anisotropy better suited the conditions for ferromagnetic resonance, and the permeability increased with increasing anisotropy. In the intermediate region, there was a non-monotonic dependence of impedance on anisotropy (and stress). The evolution of MI behavior under stress for sample No3 with an axial anisotropy is shown in Figure 14.

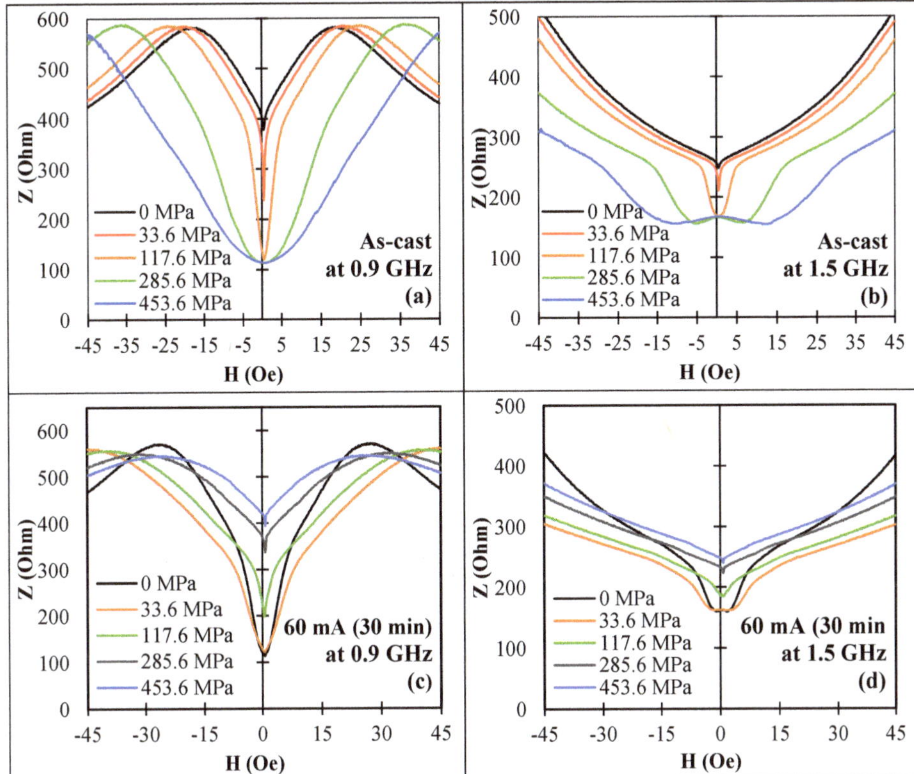

Figure 12. Effect of a tensile stress on MI vs. magnetic field for amorphous $Co_{71}Fe_5B_{11}Si_{10}Cr_3$ microwires (sample No2) in as-cast state in (**a,b**) and after current annealing (60 mA, 30 min) in (**c,d**) for different frequencies: (**a,c**) -0.9 GHz; (**b,d**) -1.5 GHz. Z corresponds to the impedance amplitude.

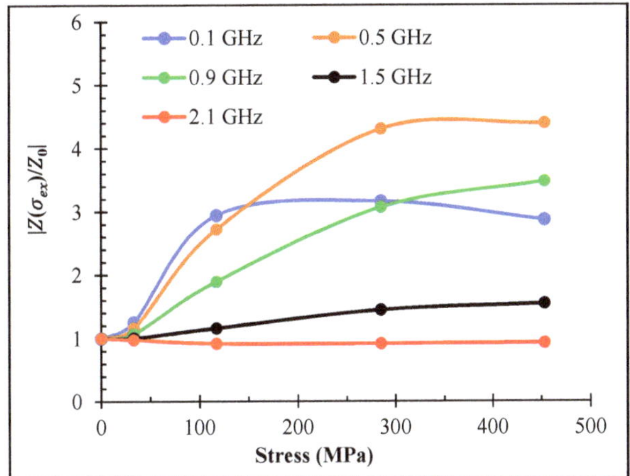

Figure 13. Stress dependence of MI at zero field $Z(H = 0)$ for amorphous $Co_{71}Fe_5B_{11}Si_{10}Cr_3$ microwires (sample No2) after DC current annealing at 60 mA for 30 min at different frequencies. $|Z(\sigma_{ex})/Z_0|=|Z(H = 0, \sigma_{ex})/Z(H = 0, \sigma_{ex} = 0)|$. σ_{ex} is the applied stress.

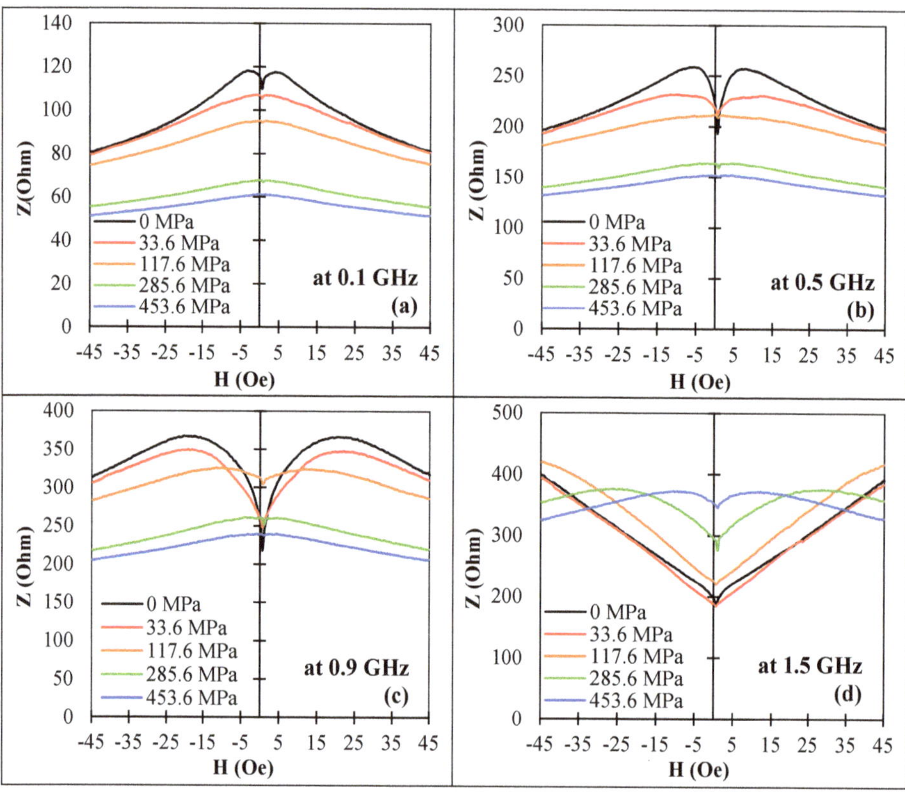

Figure 14. Effect of a tensile stress on the MI vs. magnetic field behavior for amorphous $Co_{71}Fe_5B_{11}Si_{10}Cr_3$ microwires (sample No3) in as-prepared state at frequencies: (**a**) 0.1 GHz, (**b**) 0.5 GHz, (**c**) 0.9 GHz, and (**d**) 1.5 GHz. Z corresponds to the impedance amplitude.

Axial magnetoelastic anisotropy for sample No3 was increased by increasing the annealing current, as follows from the impedance behavior at low frequencies (<100 MHz): the impedance value at zero field decreased with increasing I_{an}. This could have been due to increase in the magnetostriction constant by annealing. For $I_{an} = 90$ mA, a monotonic increase in the impedance (at zero field) was observed up to frequencies of few GHz when a tensile stress was applied, as shown in Figure 15.

Figure 16 compares the zero-field impedance vs. stress for sample No3 after different annealing conditions.

It is seen that with increasing the annealing current, the impedance stress sensitivity dropped at lower frequencies, but it increased at higher frequencies. Thus, for a wire annealed with a current of 90 mA, the impedance increased by about 300% when a stress of 450 MPa was applied.

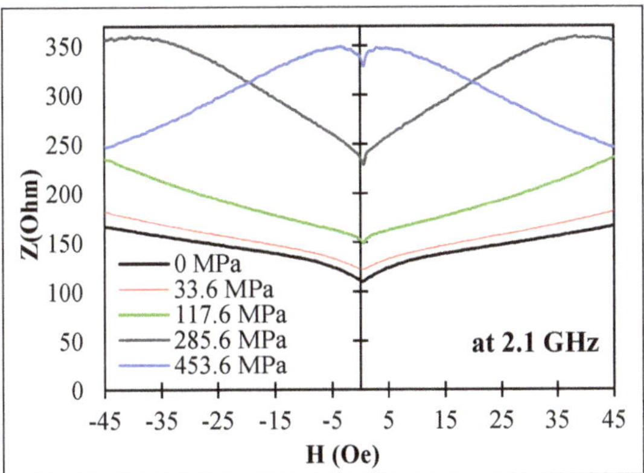

Figure 15. Effect of a tensile stress on the MI vs. magnetic field behavior for amorphous $Co_{71}Fe_5B_{11}Si_{10}Cr_3$ microwires (sample No3) after DC current annealing at 90 mA for 30 min at a frequency of 2.1 GHz. Z corresponds to the impedance amplitude.

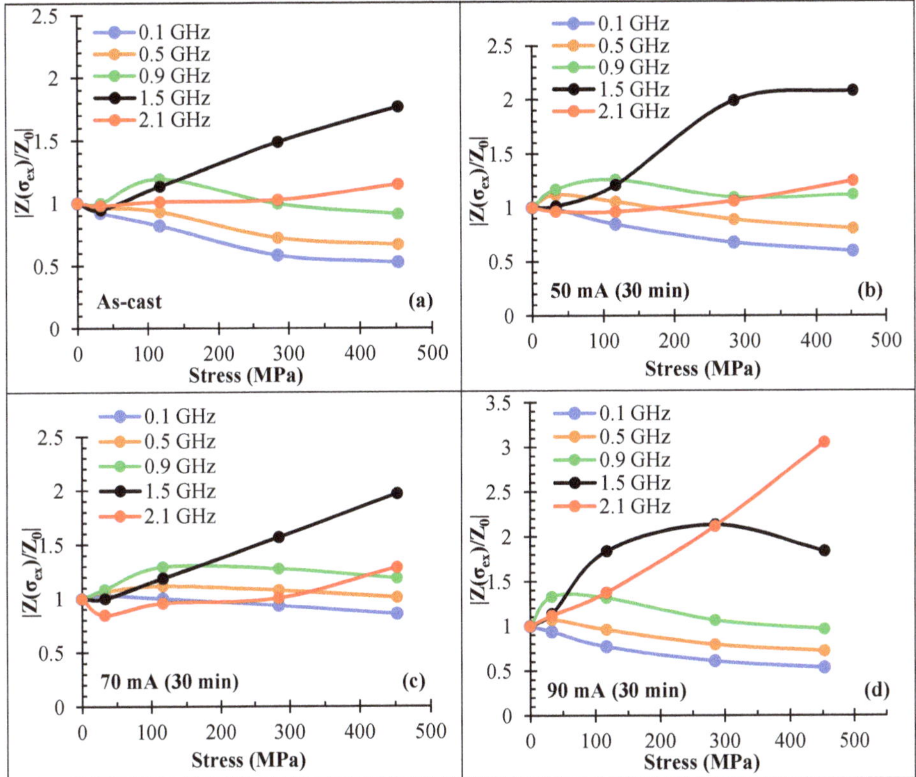

Figure 16. Stress dependence of MI at zero field ($Z(H = 0)$) for amorphous $Co_{71}Fe_5B_{11}Si_{10}Cr_3$ microwires (sample No3) in as-prepared state (**a**) and after DC current annealing at 50 (**b**), 70 (**c**), and 90 (**d**) mA for 30 min at different frequencies. $|Z(\sigma_{ex})/Z_0|=|Z(H = 0, \sigma_{ex})/Z(H = 0, \sigma_{ex} = 0)|$. σ_{ex} is the applied stress.

4. Conclusions

In this work, the MI effect at elevated frequencies was compared for Co-based microwires with different anisotropy and magnetostriction. To realize various combinations of the easy anisotropy and magnetostriction, we successfully used the current annealing treatment of amorphous and partially crystalline wires. Two mechanisms of large MI sensitivity at GHz frequencies were demonstrated: (i) due to DC magnetization re-orientation and (ii) due to anisotropy-dependent permeability dispersion. The formation of a large magnetoelastic anisotropy with an axial easy axis in current annealed Co-based microwires with partial crystallization resulted in large and sensitive stress-MI at GHz frequencies. The latter has a potential for developing stress-sensitive elements, especially for wireless operation at microwave frequencies.

The modified method of SOLT calibration with specially designed strip cells on PCB board could allow for the customized and precise impedance measurements at GHz frequency range, which is quite useful for designing various remote sensors and sensing materials operating at microwave band.

Author Contributions: Conceptualization, L.P.; methodology, J.A. and M.N.; validation, M.N., J.A. and N.Y.; formal analysis, S.P.; investigation, J.A. and M.N.; writing—original draft preparation, J.A., and M.N.; writing—review and editing, L.P. All authors have read and agreed to the published version of the manuscript.

Funding: This work was supported by the Russian Foundation for Basic Research (RFBR project no. 20-32-90129) and also by the grant from the Ministry of Education and Science of the Russian Federation in the framework of increase Competitiveness Program of NUST "MISIS". Moreover, the work was partially supported by Ministry of Science and Higher Education of the Russian Federation in the framework of the State Task (project code 0718-2020-0037).

Institutional Review Board Statement: For this type of Manuscript a formal IRB approval is not required.

Informed Consent Statement: The authors declare no conflict of interest. The funders had no role in the design of the study; in the collection, analyses, or interpretation of data; in the writing of the manuscript, or in the decision to publish the result.

Data Availability Statement: The authors confirm that the data supporting the findings of this study are available within the article. Raw data were generated at differential scanning calorimetry, B-loop meter and Hewlett-Packard 8753E Vector Network Analyzer. Derived data supporting the findings of this study are available from the corresponding author [L.P.] on request.

Acknowledgments: The authors are thankful to the MFTI ltd (http://www.microwires.com) and personally to Vladimir Larin for the donation of the wire samples used in this work.

Conflicts of Interest: The authors declare no conflict of interest.

References

1. Vázquez, M.; Chiriac, H.; Zhukov, A.; Panina, L.; Uchiyama, T. On the state-of-the-art in magnetic microwires and expected trends for scientific and technological studies. *Phys. Status Solidi Appl. Mater. Sci.* **2011**. [CrossRef]
2. Zhukova, V.; Corte-Leon, P.; González-Legarreta, L.; Talaat, A.; Blanco, J.M.; Ipatov, M.; Olivera, J.; Zhukov, A. Review of Domain Wall Dynamics Engineering in Magnetic Microwires. *Nanomaterials* **2020**, *10*, 2407. [CrossRef]
3. Alam, J.; Bran, C.; Chiriac, H.; Lupu, N.; Óvári, T.A.; Panina, L.V.; Rodionova, V.; Varga, R.; Vazquez, M.; Zhukov, A. Cylindrical micro and nanowires: Fabrication, properties and applications. *J. Magn. Magn. Mater.* **2020**. [CrossRef]
4. Baca, A.J.; Ahn, J.H.; Sun, Y.; Meitl, M.A.; Menard, E.; Kim, H.S.; Choi, W.M.; Kim, D.H.; Huang, Y.; Rogers, J.A. Semiconductor wires and ribbons for high-performance flexible electronics. *Angew. Chem. Int. Ed.* **2008**. [CrossRef]
5. Moya, A.; Archilla, D.; Navarro, E.; Hernando, A.; Marín, P. Scattering of microwaves by a passive array antenna based on amorphous ferromagnetic microwires for wireless sensors with biomedical applications. *Sensors* **2019**, *19*, 3060. [CrossRef]
6. Qin, F.; Peng, H.X. Ferromagnetic microwires enabled multifunctional composite materials. *Prog. Mater. Sci.* **2013**. [CrossRef]
7. Mohri, K.; Uchiyama, T.; Shen, L.P.; Cai, C.M.; Panina, L.V. Amorphous wire and CMOS IC-based sensitive micro magnetic sensors (MI sensor and SI sensor) for intelligent measurements and controls. *J. Magn. Magn. Mater.* **2002**. [CrossRef]
8. Gudoshnikov, S.; Usov, N.; Nozdrin, A.; Ipatov, M.; Zhukov, A.; Zhukova, V. Highly sensitive magnetometer based on the off-diagonal GMI effect in Co-rich glass-coated microwire. *Phys. Status Solidi Appl. Mater. Sci.* **2014**. [CrossRef]

Dolabdjian, C.; Ménard, D. Giant magneto-impedance (GMI) magnetometers. In *Smart Sensors, Measurement and Instrumentation*; Springer: Cham, Switzerland, 2017. [CrossRef]

Uchiyama, T.; Takiya, T. Development of precise off-diagonal magnetoimpedance gradiometer for magnetocardiography. *AIP Adv.* **2017**. [CrossRef]

Zhukova, V.; Corte-Leon, P.; Ipatov, M.; Blanco, J.M.; Gonzalez-Legarreta, L.; Zhukov, A. Development of magnetic microwires for magnetic sensor applications. *Sensors* **2019**, *19*, 4767. [CrossRef] [PubMed]

Makhnovskiy, D.P.; Panina, L.V.; Mapps, D.J. Field-dependent surface impedance tensor in amorphous wires with two types of magnetic anisotropy: Helical and circumferential. *Phys. Rev. B Condens. Matter Mater. Phys.* **2001**. [CrossRef]

Panina, L.V.; Makhnovskiy, D.P.; Morchenko, A.T.; Kostishin, V.G. Tunable permeability of magnetic wires at microwaves. *J. Magn. Magn. Mater.* **2015**. [CrossRef]

Ciureanu, P.; Britel, M.; Ménard, D.; Yelon, A.; Akyel, C.; Rouabhi, M.; Cochrane, R.W.; Rudkowski, P.; Ström-Olsen, J.O. High frequency behavior of soft magnetic wires using the giant magnetoimpedance effect. *J. Appl. Phys.* **1998**, *83*, 6563–6565. [CrossRef]

Da Silva, R.B.; De Andrade, A.M.H.; Severino, A.M.; Schelp, L.F.; Sommer, R.L. Giant magnetoimpedance in glass-covered amorphous microwires at microwave frequencies. *J. Appl. Phys.* **2002**. [CrossRef]

Zhao, Y.; Wang, Y.; Estevez, D.; Qin, F.; Wang, H.; Zheng, X.; Makhnovskiy, D.; Peng, H. Novel broadband measurement technique on PCB cells for the field- And stress-dependent impedance in ferromagnetic wires. *Meas. Sci. Technol.* **2020**. [CrossRef]

Kraus, L.; Knobel, M.; Kane, S.N.; Chiriac, H. Influence of Joule heating on magnetostriction and giant magnetoimpedance effect in a glass covered CoFeSiB microwire. *J. Appl. Phys.* **1999**. [CrossRef]

Liu, J.S.; Cao, F.Y.; Xing, D.W.; Zhang, L.Y.; Qin, F.X.; Peng, H.X.; Xue, X.; Sun, J.F. Enhancing GMI properties of melt-extracted Co-based amorphous wires by twin-zone Joule annealing. *J. Alloys Compd.* **2012**. [CrossRef]

Zhukov, A.; Talaat, A.; Ipatov, M.; Blanco, J.M.; Zhukova, V. Tailoring of magnetic properties and GMI effect of Co-rich amorphous microwires by heat treatment. *J. Alloys Compd.* **2014**. [CrossRef]

Thiabgoh, O.; Shen, H.; Eggers, T.; Galati, A.; Jiang, S.; Liu, J.S.; Li, Z.; Sun, J.F.; Srikanth, H.; Phan, M.H. Enhanced high-frequency magneto-impedance response of melt-extracted Co69.25Fe4.25Si13B13.5 microwires subject to Joule annealing. *J. Sci. Adv. Mater. Devices* **2016**. [CrossRef]

Nematov, M.G.; Baraban, I.; Yudanov, N.A.; Rodionova, V.; Qin, F.X.; Peng, H.X.; Panina, L.V. Evolution of the magnetic anisotropy and magnetostriction in Co-based amorphous alloys microwires due to current annealing and stress-sensory applications. *J. Alloys Compd.* **2020**. [CrossRef]

Panina, L.V.; Sandacci, S.I.; Makhnovskiy, D.P. Stress effect on magnetoimpedance in amorphous wires at gigahertz frequencies and application to stress-tunable microwave composite materials. *J. Appl. Phys.* **2005**. [CrossRef]

Adenot, A.L.; Acher, O.; Pain, D.; Duverger, F.; Malliavin, M.J.; Damiani, D.; Taffary, T. Broadband permeability measurement of ferromagnetic thin films or microwires by a coaxial line perturbation method. *J. Appl. Phys.* **2000**. [CrossRef]

Melo, L.G.C.; Ciureanu, P.; Yelon, A. Permeability deduced from impedance measurements at microwave frequencies. *J. Magn. Magn. Mater.* **2002**. [CrossRef]

Rumiantsev, A.; Ridler, N. VNA calibration. *IEEE Microw. Mag.* **2008**. [CrossRef]

Kilinc, M.; Garcia, C.; Eginligil, M.; Wang, J.; Huang, W. De-embedding zero-field signal in high-frequency magneto-impedance measurements of soft ferromagnetic materials. *J. Magn. Magn. Mater.* **2019**. [CrossRef]

Chiriac, H.; Corodeanu, S.; Lostun, M.; Stoian, G.; Ababei, G.; Óvári, T.A. Rapidly solidified amorphous nanowires. *J. Appl. Phys.* **2011**. [CrossRef]

Baranov, S.A.; Larin, V.S.; Torcunov, A.V. Technology, preparation and properties of the cast glass-coated magnetic microwires. *Crystals* **2017**, *7*, 136. [CrossRef]

Nematov, M.G.; Salem, M.M.; Adam, A.M.; Ahmad, M.; Yudanov, N.A.; Panina, L.V.; Morchenko, A.T. Effect of Stress on Magnetic Properties of Annealed Glass-Coated Co71Fe5B11Si10Cr3 Amorphous Microwires. *IEEE Trans. Magn.* **2017**. [CrossRef]

Dzhumazoda, A.; Panina, L.V.; Nematov, M.G.; Ukhasov, A.A.; Yudanov, N.A.; Morchenko, A.T.; Qin, F.X. Temperature-stable magnetoimpedance (MI) of current-annealed Co-based amorphous microwires. *J. Magn. Magn. Mater.* **2019**. [CrossRef]

Zhukov, A.; Churyukanova, M.; Kaloshkin, S.; Sudarchikova, V.; Gudoshnikov, S.; Ipatov, M.; Talaat, A.; Blanco, J.M.; Zhukova, V. Magnetostriction of Co-Fe-based amorphous soft magnetic microwires. In *Energy Technology 2015: Carbon Dioxide Management and Other Technologies*; Springer: Cham, Switzerland, 2016. [CrossRef]

Kostitsyna, E.V.; Gudoshnikov, S.A.; Popova, A.V.; Petrzhik, M.I.; Tarasov, V.P.; Usov, N.A.; Ignatov, A.S. Mechanical properties and internal quenching stresses in Co-rich amorphous ferromagnetic microwires. *J. Alloys Compd.* **2017**. [CrossRef]

Fitzpatrick, J. Error models for system measurement. *Microw. J.* **1978**, *21*, 63–66.

Salem, M.M.; Nematov, M.G.; Uddin, A.; Panina, L.V.; Churyukanova, M.N.; Marchenko, A.T. CoFe-microwires with stress-dependent magnetostriction as embedded sensing elements. *J. Phys. Conf. Ser.* **2017**. [CrossRef]

Herzer, G. Modern soft magnets: Amorphous and nanocrystalline materials. *Acta Mater.* **2013**. [CrossRef]

Vázquez, M.; González, J.; Hernando, A. Induced magnetic anisotropy and change of the magnetostriction by current annealing in Co-based amorphous alloys. *J. Magn. Magn. Mater.* **1986**. [CrossRef]

MDPI
St. Alban-Anlage 66
4052 Basel
Switzerland
Tel. +41 61 683 77 34
Fax +41 61 302 89 18
www.mdpi.com

Nanomaterials Editorial Office
E-mail: nanomaterials@mdpi.com
www.mdpi.com/journal/nanomaterials